Fundamental Principles
of
Manufacturing Processes

Fundamental Principles
of
Manufacturing Processes

by Robert H. Todd and Dell K. Allen,
Brigham Young University

and Leo Alting,
Technical University of Denmark

Industrial Press Inc.
New York

HOUSTON PUBLIC LIBRARY

Library of Congress Cataloging-in-Publication Data
Todd, Robert, 1942–
   Fundamental principles of manufacturing processes / by Robert H. Todd,   Dell K.
Allen, Leo Alting.—1st ed.
      176 p. 19.7 × 25.4 cm.
   Includes index
   ISBN 0-8311-3050-4
   1. Manufacturing processes.   I. Allen, Dell K., 1931–   .   II. Alting,   Leo,
1939–   .   III. Title
   TS183.T58   1994
   670.42—dc20   93-49409
                                   CIP

INDUSTRIAL PRESS INC.
200 Madison Avenue
New York, New York 10016–4078

FUNDAMENTAL PRINCIPLES OF MANUFACTURING PROCESSES

First Edition, 1994

First Printing

Rights to publish copyrighted data base charts and figures that are used in this publication have been obtained from the Manufacturing Consortium now administered by the college of engineering and technology and originally developed under the direction of CAM Software Research Center at Brigham Young University. This material may not be reproduced in any form without permission.

10 9 8 7 6 5 4 3 2

Contents

Conversion Factors

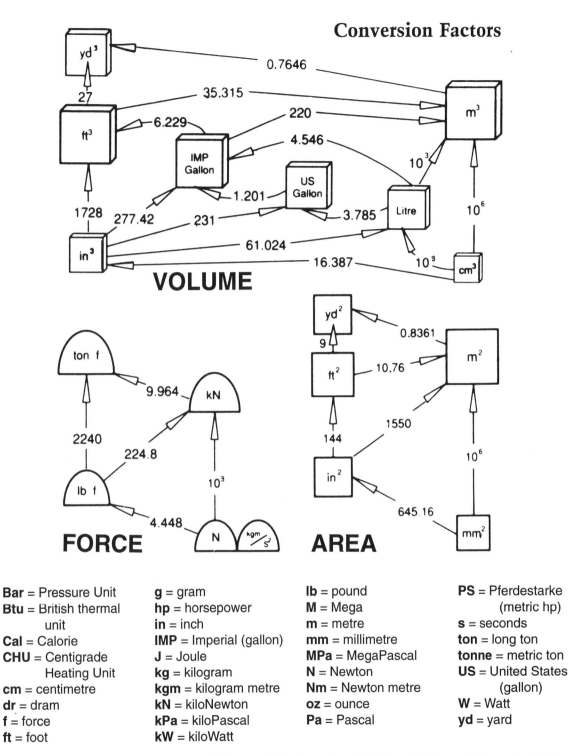

VOLUME

FORCE

AREA

Bar = Pressure Unit	**g** = gram
Btu = British thermal unit	**hp** = horsepower
	in = inch
Cal = Calorie	**IMP** = Imperial (gallon)
CHU = Centigrade Heating Unit	**J** = Joule
	kg = kilogram
cm = centimetre	**kgm** = kilogram metre
dr = dram	**kN** = kiloNewton
f = force	**kPa** = kiloPascal
ft = foot	**kW** = kiloWatt

lb = pound	**PS** = Pferdestarke (metric hp)
M = Mega	
m = metre	**s** = seconds
mm = millimetre	**ton** = long ton
MPa = MegaPascal	**tonne** = metric ton
N = Newton	**US** = United States (gallon)
Nm = Newton metre	
oz = ounce	**W** = Watt
Pa = Pascal	**yd** = yard

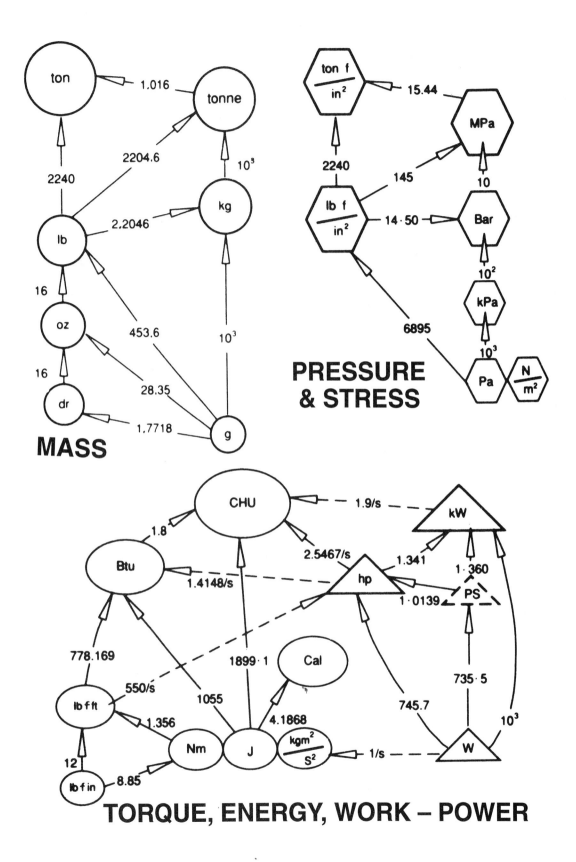

MASS

PRESSURE & STRESS

TORQUE, ENERGY, WORK – POWER

General Structure of Manufacturing Processes

1.1 Introduction

The importance of improving manufacturing processes grows each year. Manufacturing production is central to any manufacturing enterprise. With the rapid advancements in electronics and computer technology, manufacturing production systems are becoming increasingly computer-integrated. These systems are powerful competitive weapons because they enable industry to produce products with low response time, low inventory, and flexibility. There is, however, a long way to go before computer-controlled production systems become a complete reality.

One of the most important factors in improving manufacturing and developing automated production systems is a thorough knowledge of manufacturing production processes, including data about the processes and conditions under which the processes are carried out.

In manufacturing, the actual materials and equipment used are costly, but these costs are substantially determined by those responsible for product design before manufacturing even begins. By virtue of decisions made early in the production process, designers determine up to 70% of manufacturing costs. Lacking sufficient knowledge and experience, designers may make poor decisions about materials, tolerances, shapes, size, and product function. All of these factors have a tremendous impact on the processes used in manufacturing the component parts of any product.

Thus, a thorough knowledge of manufacturing processes is necessary not only for individuals involved in the actual making of things, but also for those who are involved in the design phase. Manufacturing is an interdisciplinary activity. It is vital that individuals in all phases of the manufacturing enterprise consider decisions about manufacturing processes rather than considering only one particular aspect such as marketing, design, or other aspects of the product development process. Many important and costly decisions are made before the manufacturing process even begins, and it is these decisions that most drastically affect product cost, quality, quickness to market, and customer satisfaction. If better decisions can be made in the early phases of product design, better products will result for customers, and greater profits will result for manufacturing companies.

In the industrial environment, over 300 individual manufacturing processes have been identified, and the number continues to grow. It thus takes considerable time for an individual to acquire even a cursory knowledge about this vast array of existing processes. From an educational point of view, it is not possible to teach all the processes to a student. From an applicational point of view, it is difficult to acquire a survey of processing possibilities. To overcome these difficulties, a systematic approach to manufacturing has been adopted. This systematic approach can be characterized by grouping the processes into families. Within each family several common characteristics exist. By studying the characteristics of each process family, a thorough foundation can be more easily obtained and assimilated. Any new knowledge about specific processes then may be added to this foundation.

1.2 Definition of the Process Families

An analysis of manufacturing processes leads to certain common characteristics that can be formulated into a model of manufacturing processes. One such model, developed by Alting, is very useful. This model includes material flow, information flow, and energy flow systems. In the context of this book, only the work material flow system will be discussed.

The work material flow system is described by the following three characteristics:

* Type of process
* State of the workpiece material
* Nature of the processing energy

Type of Process

Two major divisions of processes can be distinguished. *Shaping* processes change the basic geometry, or shape, of a workpiece. *Nonshaping* or *treating* processes change the properties of the workpiece material. The type of process is identified according to the function of the process.

Shaping processes are divided into three categories:

* In **mass-conserving processes**, the mass of the workpiece material before the process is equal to the mass after; in other words, the change in mass is zero. This can be represented as $dM = 0$, where dM represents the change in mass of material.
* In **mass-reducing processes**, the desired geometry is produced by removing material from the initial piece of work material (the final component can be circumscribed by the initial piece), and $dM < 0$.
* In **joining processes**, two or more components are joined permanently to produce a new component, and $dM > 0$.

State of the Workpiece Material

Within each of the three basic types of shaping processes, the workpiece material can exist in different physical states when the actual shaping is performed. The workpiece material can be processed in a solid, liquid, granular, or vapor state. In each state, a different processing energy can be used to shape the material.

Nature of the Processing Energy

For each type of shaping process, and for a given state of workpiece material, only certain physical methods of processing the workpiece material (process energies) can be used. The basic methods of shaping materials, or the nature of interaction between the process and the workpiece, are as follows:

* Mechanical (i.e., plastic deformation, fracture, flow)
* Thermal (i.e., heating, melting, evaporation)
* Chemical (i.e., solution, dissolution)

Normally, the performance of a basic shaping process can be divided into the following three stages:

1. Preparation of the workpiece material
2. Shaping of the workpiece material
3. Stabilization of the workpiece material shape

When we use the term "basic process" in this context, it will, unless otherwise stated, concern the actual shaping method. In casting, for example, preparation may involve melting the workpiece material. Stabilization of the casting may involve solidification of the workpiece material after mechanically filling the die cavity with the molten material when shaping of the workpiece is carried out.

Based on the three elements of the workpiece material flow system, appropriate basic process grouping can be established. Nonshaping processes modify the engineering and the aesthetic properties of materials. For example, a heat treatment modifies the surface properties of a material. These processes are classified according to function.

Table 1–1 lists the 10 process families described in this book. There are six **shaping** process families (see Table 1–1). Based on the joint efforts of Allen and Alting, these families have been formed to create a process taxonomy of all known shaping processes.

Nonshaping processes can be divided into four process families according to function (see lower portion of Table 1–1).

The taxonomy that follows shows the process families and the individual processes that belong to each family. This classification, or taxonomy, is now

Table 1–1. Classification of processes into 10 families according to state of material and nature of processing energy

Type of process (family)	State of material	Nature of process energy
Shaping		
1. Mass-reducing	Solid	Mechanical
2. Mass-reducing	Solid	Thermal
3. Mass-reducing	Solid	Chemical
4. Mass-conserving	Solid/granular	Mechanical
5. Consolidation	Liquid/plastic	Mechanical
6. Joining	Solid (except adjacent surfaces)	Mechanical
Nonshaping		
7. Hardening	Solid	Chemical/thermal
8. Softening	Solid	Chemical/thermal
9. Surface treatment	Preparation	Mechanical/thermal/chemical
10. Surface treatment	Coating	Mechanical/thermal/chemical

used in the United States and Europe. The taxonomy is a valuable tool in identifying processes and their capabilities.

This textbook discusses the 10 process families, based on their common characteristics. Individual processes are described in an accompanying *Reference Guide*, which provides information for understanding and applying each of the selected processes. The description of the processes is carried out according to a carefully developed format. As shown by the taxonomy, only a portion of the total number of known processes has been selected to be included in the *Reference Guide*. The process selection is, to a large extent, based on industrial priorities as suggested by the Manufacturing Consortium Industrial Members who have guided the project.

Taxonomy of Manufacturing Processes

The Manufacturing Processes Taxonomy, based on the process classification system initially developed at Brigham Young University[1] and later adapted by members of the Manufacturing Consortium, provides a concise roadmap of some 300 processes used for modifying geometry or properties of engineering materials.

It has been said that students can learn twice as much in half the time when the material to be studied has been classified and the critical attributes have been clearly identified. In this text, we attempt to do both.

Processes used for modifying workpiece *geometry* are called "shaping" processes. Processes used for modifying *properties* of materials are called "nonshaping" processes. Shaping processes have been grouped into "mass-conserving" processes, "mass-reducing" processes, and mass-increasing, or "joining," processes. Nonshaping processes have been grouped into processes dealing with heat treatment and with surface finishing. Each of these processes has been further subdivided into fourteen major "families" of processes, as shown on the first sheet of the Manufacturing Processes Taxonomy chart. In turn, each of the process families has been subdivided into unique individual processes.

---

[1] Allen, Dell K. and Paul R. Smith, *Process Classification*, Monograph No. 5, Computer Aided Manufacturing Laboratory Brigham, Brigham Young University, Provo, Utah, January 1980.

Manufacturing Processes

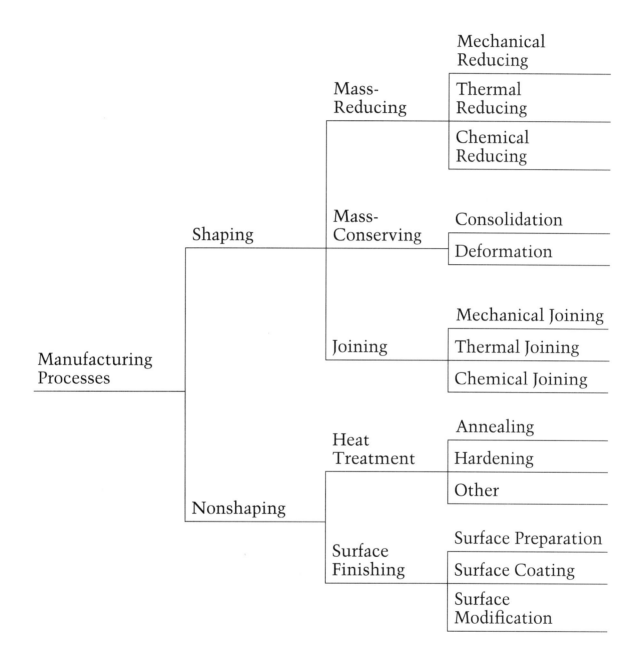

Taxonomy of Manufacturing Processes: Manufacturing Processes

Mass-Reducing

- **Mass-Reducing**
  - **Mechanical Reducing**
    - **Reducing (Chips)**
      - **Single-Point Cutting**
        - Turning/Facing
        - Boring
        - Shaping/Planing
        - Parting/Grooving
        - Threading (SP)
      - **Multipoint Cutting**
        - Drilling
        - Reaming
        - Milling/Routing
        - Broaching
        - Threading (MP)
        - Filing
        - Sawing
        - Gear Cutting
      - **Abrasive Machining**
        - Grinding
        - Honing
        - Lapping
        - Superfinishing
        - Ultrasonic Machining
        - Jet Machining
    - **Separating (Shear)**
      - **Shearing**
        - Squaring
        - Slitting
        - Rotary Shearing
        - Nibbling
      - **Blanking**
        - Conventional Blanking
        - Steel-Rule-Die Blanking
        - Fine Blanking
        - Shaving/Trimming
        - Dinking
      - **Piercing**
        - Punching
        - Perforating
        - Lancing
        - Notching

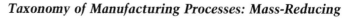

Taxonomy of Manufacturing Processes: Mass-Reducing

		Air Arc Cutting
	Torch Cutting	Gas Cutting
Thermal Reducing		Plasma Arc Cutting
	Electrical Discharge Machining	Cavity-Type EDM
		EDM Grinding
		EDM Sawing
	High Energy Beam Machining	Electron Beam Cutting
		Laser Beam Cutting
		Ion Beam Cutting
	Chemical Milling	Immersion Chemical Milling
Chemical Reducing		Spray Chemical Milling
	Electrochemical Milling	Cavity-Type ECM
		Grinder-Type ECM
	Photochemical Milling	Photo Etching
		Photo Milling

Taxonomy of Manufacturing Processes: Mass-Reducing

Mass-Conserving

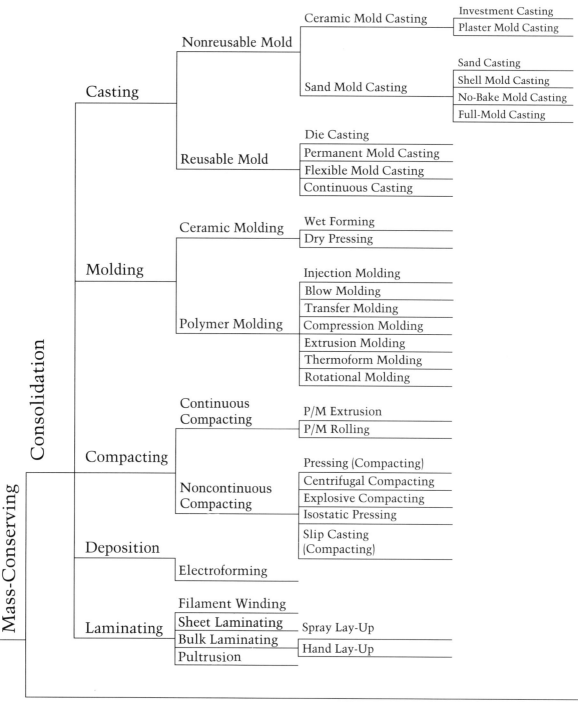

Taxonomy of Manufacturing Processes: Mass-Conserving

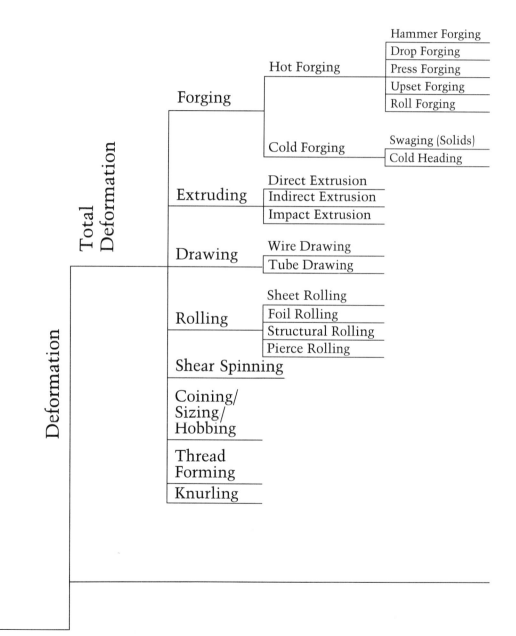

Taxonomy of Manufacturing Processes: Mass-Conserving

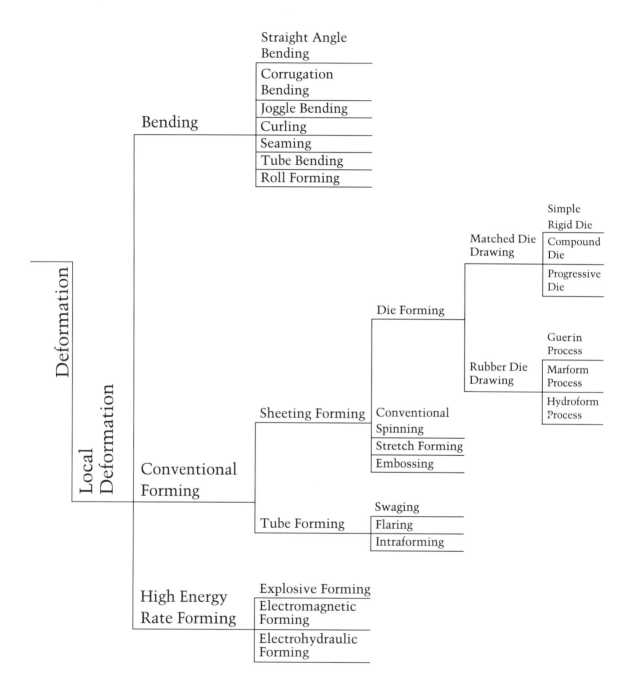

Taxonomy of Manufacturing Processes: Mass-Conserving

Joining

Mechanical Joining
- Pressure (Cold) Welding
- Friction (Inertial) Welding
- Ultrasonic Welding
- Explosive Welding

Thermal Joining

Thermal Welding

Electric Arc Welding
- Shielded Metal Arc Welding
- Gas Metal Arc (MIG) Welding
- Gas Tungsten Arc (TIG) Welding
- Submerged Arc Welding
- Carbon Arc Welding
- Stud Welding

Electrical Resistance Welding
- Spot Welding
- Seam Welding
- Projection Welding
- Butt Welding
- Percussion Welding
- Electroslag Welding

Gas/Chemical Welding
- Combustible Gas Welding
- Atomic Hydrogen Welding

Braze Welding
- Gas Brazing
- Carbon Arc Brazing

Diffusion Bonding

High Energy Beam Welding
- Electron Beam Welding
- Laser Beam Welding
- Plasma Arc Welding

Chemical Joining

Brazing
- Infrared Brazing
- Resistance Brazing
- Torch Brazing
- Dip Brazing
- Furnace Brazing
- Induction Brazing

Soldering
- Friction/Ultrasonic Soldering
- Induction Soldering
- Infrared Soldering
- Dip Soldering
- Iron Soldering
- Resistance Soldering
- Torch Soldering
- Wave Soldering

Adhesive Bonding

Taxonomy of Manufacturing Processes: Joining

Heat Treatment

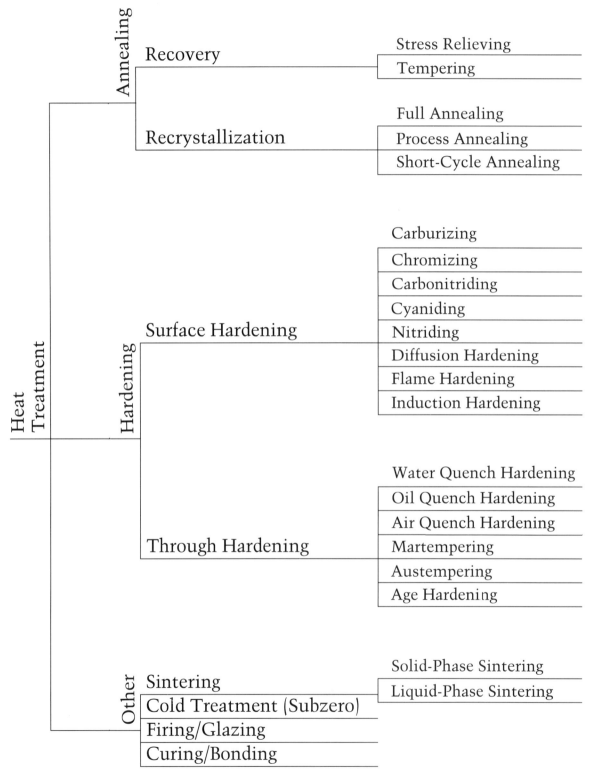

Taxonomy of Manufacturing Processes: Heat Treatment

Surface Finishing

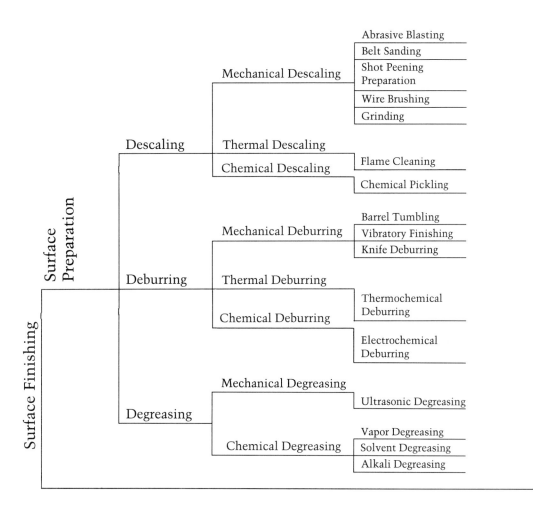

Taxonomy of Manufacturing Processes: Surface Finishing

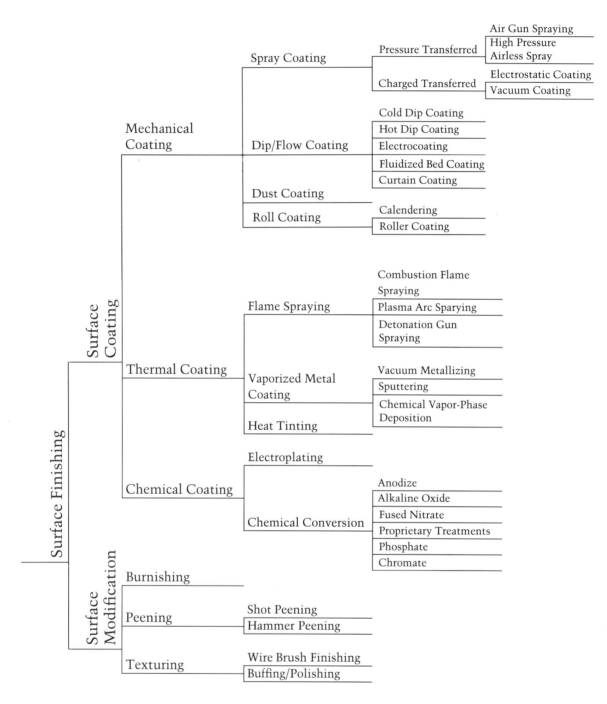

Taxonomy of Manufacturing Processes: Surface Finishing

Mechanical Mass-Reducing Processes

2.1 Introduction

Mass-reducing processes (see Chapter 1, the Taxonomy) include all processes in which the desired geometry is created by removing excess material from a solid workpiece. Mass-reducing processes can be classified into three categories according to how the excess material is removed: mechanical, thermal, and chemical. Mechanical mass-removal processes include those processes that are sometimes classified as finishing processes and that only remove a small amount of material.

Mechanical mass-reducing processes generally have the following characteristics:

* The desired shape is obtained by removing material from a solid workpiece
* Material is removed mechanically through a controlled fracture by means of cutting, abrasion, or shearing
* Creation of the desired workpiece shape results from relative motions between a solid workpiece and a tool
* Tools used are usually rigid, but they can also be in the form of abrasive particles or a liquid jet stream
* The properties of the solid workpiece allow processing by cutting, abrasion, or shearing

2.2 Types of Mechanical Processes

Mechanical mass-reducing uses controlled fracture to remove material from a solid stock. Within this category, the processes can be divided into two sub-categories:

* Reducing (chips)
* Separating (shear)

Each of these two methods is governed by its own set of rules. The basic differences between these methods will be introduced here, and the remaining sections of this chapter will discuss each method in detail.

Reducing (Chips)

This category includes all mechanical mass-removal processes that remove material from the workpiece in the form of chips. The removal tool and the workpiece are moved with respect to each other so that small chips are sheared away from the workpiece until the desired shape is obtained. Chip removal may be done in the following three ways:

* Single-point cutting
* Multipoint cutting
* Abrasive machining

Single-point cutting processes use a tool with a single cutting edge. These processes include turning, boring (see Figure 2–1), and parting.

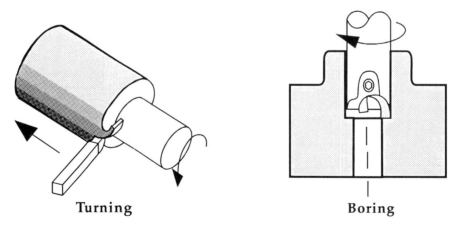

Turning   Boring

Figure 2–1. Examples of reducing processes (single-point cutting).

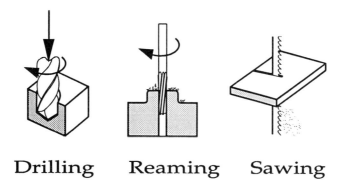

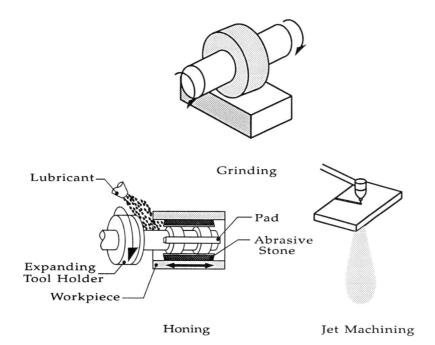

Figure 2–2. Examples of reducing processes (multipoint cutting).

Multipoint cutting processes use a multiedge tool, such as those used for drilling, reaming, and sawing (see Figure 2–2).

Abrasion machining processes use tiny abrasive particles to remove chips from the workpiece. These abrasive particles are either guided through a jet stream or bonded together in a rigid tool, depending on the process used. During abrasive processes, many particles remove small chips from the workpiece simultaneously. Examples of abrasive machining processes are grinding, honing, and jet machining (see Figure 2–3).

Figure 2–3. Examples of reducing processes (abrasive machining).

Separating (Shear)

In separating processes, larger portions of the workpiece are "separated" from the workpiece. Opposing motions between upper and lower tools create a narrow shearing zone that fractures the workpiece. Common examples of separating are shearing, blanking, and punching (see Figure 2–4).

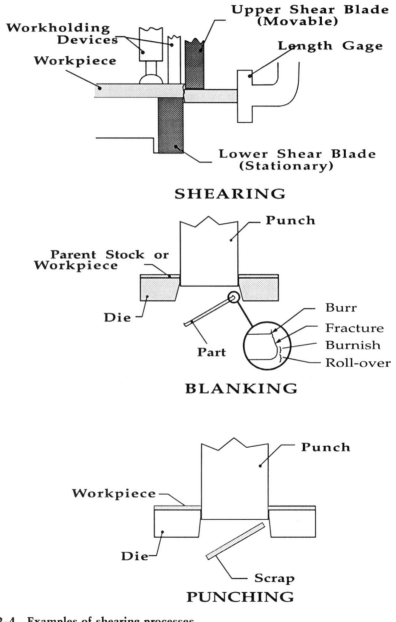

Figure 2–4. Examples of shearing processes.

2.3 Reducing (Cutting Processes)

As mentioned before, the reducing processes, or chip removal processes, can be divided into the three following groups:

* Single-point cutting
* Multipoint cutting
* Abrasive machining

Since the theory behind all cutting operations is similar, this section will discuss both single-point and multipoint processes. Section 2.4 will discuss abrasive machining processes.

General Definition of Cutting Processes

In general, cutting processes have the following characteristics:

* Chip removal is based on controlled fracture, which is created by relative motions between a shaped, rigid tool and a workpiece
* The cutting tool is rigid, has a well-defined shape, and can be classified as single-point or multipoint
* Motions between the tool and the workpiece are either translational, rotational, or a combination of the two

To better evaluate the capabilities of cutting processes the following topics will be discussed: process mechanics, tool types, tool materials, process conditions, workpiece materials, and cutting process expectations.

Process Mechanics of Cutting

Cutting is based on controlled fracture. The material is subjected to a narrow zone of shearing, which leads to fracture. Figure 2–5 shows the general cutting principle.

In general, chip formation (see Figure 2–9) is influenced by the following factors:

* Tool geometry
* Tool material
* Workpiece material type (including material properties)
* Cutting conditions
* Coolant/lubrication
* Cutting machine (static/dynamic behavior, tool holding/work holding)

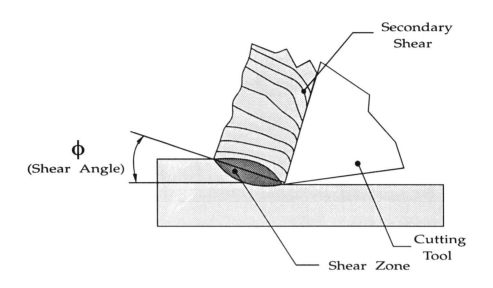

Figure 2–5. Cutting principle.

Figure 2–6 shows typical tool geometries for turning and drilling, with definitions of faces and angles. It also shows definitions of cutting conditions for turning.

Figure 2–7 shows conditions for plane milling and end milling.

Process mechanics can be better understood by using Figure 2–8. In this figure, one can see that *shear deformation* is confined to a shear plane defined by the shear angle. In practice, shearing is not confined to a plane, but rather to a narrow *shear zone*. This zone is actually a group of shear planes lined up along the theoretical shear plane. At low cutting speeds, the shear zone becomes thicker, but at high cutting speeds, it becomes very narrow.

Several values and formulas are commonly used to understand cutting mechanics better. The first of these is the cutting ratio or chip thickness ratio. This ratio, r_c is defined below.

$$r_c = \frac{h_1}{h_2} \qquad (2\text{–}1)$$

h_1 = depth of cut
h_2 = chip thickness

The chip thickness ratio, r_c can be used as an indicator of the quality of the cutting process. Another commonly used value is the inverse chip, or chip com-

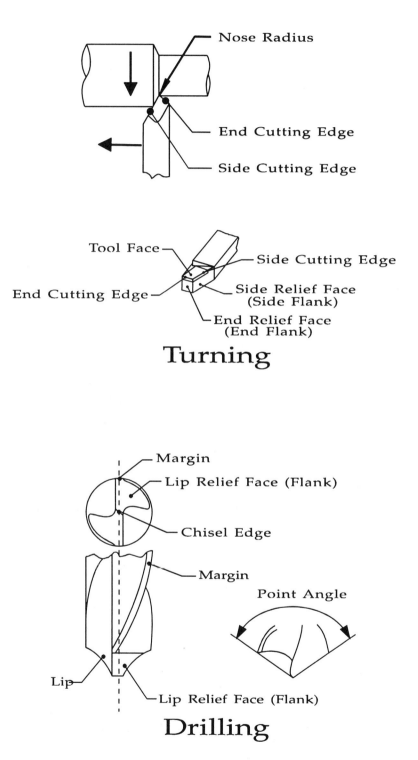

Figure 2–6. Definitions for turning and drilling; definitions of faces and angles.

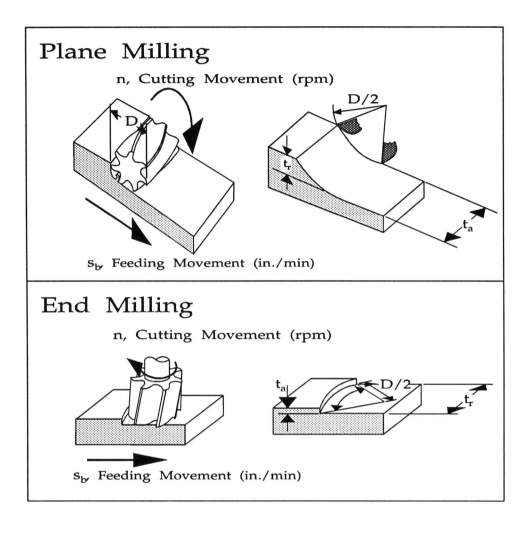

Feed (for table), in./min	s_b
Feed (per rev. of cutter), in./rev.	$s_n = s_b/n$
Cutting speed	$V = \pi D n$
Removal rate, U in.^{3}/min	$U = t_r \cdot t_a \cdot s_b$

Figure 2–7. **Conditions for plane milling and end milling.**

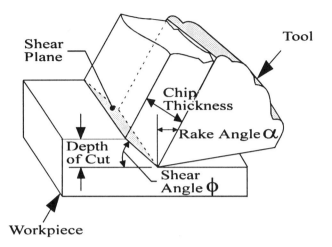

Figure 2–8. Geometry and angles in orthogonal cutting.

pression, ratio. This ratio, h, is the inverse of the chip thickness ratio ($h = 1/r_c$) and can be used along with the rake angle (α) to determine the shear angle (ϕ).

$$\tan \phi = \frac{\cos \alpha}{h - \sin \alpha} \qquad (2\text{–}2)$$

α = rake angle
ϕ = shear zone angle
h = inverse chip ratio

As the shear angle, ϕ, is decreased, chip thickness, h_2, and shear zone length are increased, which also increases force and power requirements. The actual shear angle can be ascertained experimentally by measuring h_2 and using the value obtained from (2–1) and (2–2).

Cutting process quality can be manipulated by adjusting certain factors. A larger shear angle will reduce the power requirements to remove a given amount of material in a given amount of time. The chip compression ratio ($h = 1/r_c$) should be kept as small as possible. Decreasing the chip compression ratio increases the shear angle, which in turn decreases power consumption. Hard materials give lower chip compression values than soft materials, but they require higher overall cutting forces. Friction between the tool and workpiece increases chip compression; it also increases the required cutting force. This friction can be reduced by introducing a suitable cutting fluid between the tool and workpiece or by utilizing free-machining materials such as leaded steels. The chip compression can be

Discontinuous   Continuous   Built-up Edge

Figure 2–9. Types of chips.

reduced further by increasing the cutting speed or the feed, but an upper limit is set by the acceptable tool life.

The type of chip formed provides valuable information on the efficiency of the cutting process and is determined by the properties of the work material, the geometry of the cutting tool, and the cutting conditions. Figure 2–9 shows the three general types of chips: discontinuous (segmental) chips, continuous chips, and continuous chips with built-up edges.

Discontinuous chips usually form from the cutting of brittle materials like cast iron and brass; the result is a fairly good surface finish. Under certain conditions, discontinuous chips can also be produced in more ductile materials like steel. Low cutting speeds or low rake angles can produce discontinuous chips from ductile materials; this results in rough surfaces and poor tool life. An increase in rake angle or cutting speed normally changes discontinuous chips into more desirable continuous chips.

Continuous chips characteristically form when cutting most ductile materials. Sometimes continuous long chips may be difficult to handle, so a tool with a chip breaker is used to curl and break the chips into short lengths. The chip breaker is usually molded or ground into the cutting tool inserts.

Cutting ductile materials at low speeds with high friction sometimes causes some of the chip material to adhere to the tool face. This forms a built-up edge, which creates a rough surface. Increasing the cutting speed usually solves this problem.

Types of Cutting Tools

There are two main groups of cutting tools:

* Single-point tools with one major cutting edge are used for turning, boring, shaping, planing, parting, grooving, and threading

* Multipoint tools with more than one cutting edge are used for drilling, milling, reaming, sawing, broaching, gear cutting, threading, and filing

Both groups have well-defined and highly standardized edge geometries, as shown in Figure 2–10.

Selecting tool type and edge geometry depends on the work material, the state of the machine tool, and the cutting conditions. In general, both single-point and multipoint cutting tools are bought as standard tools from a supplier, so it is necessary to know what is on the market.

Cutting Tool Materials

In general, the following materials are used for cutting tools:

* High speed steels (HSS)
* Cemented carbides (uncoated and coated)
* Ceramics
* Diamonds
* Cubic boron nitride

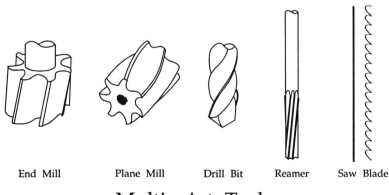

Turning Tool

Single-Point Tool

End Mill Plane Mill Drill Bit Reamer Saw Blade

Multipoint Tools

Figure 2–10. Classification of cutting tools.

High speed steels are alloyed steels that permit a cutting edge temperature in the range of 500–600°C (900–1000°F). High speed steels are used extensively in twist drills, milling cutters, and special purpose tools. Cemented carbides in the form of replaceable inserts take over many of the applications of solid HSS tools. Carbides can be operated at higher speeds and are often less expensive than high speed steel tools because of the absence of tool regrinding.

Cemented/sintered carbides are carbides such as WC (tungsten carbide) and TiC (titanium carbide) that are produced in powder metallurgic processes with cobalt as the binder. A large variety of sintered carbides exists, and each is generally developed to cut different material groups effectively. They are normally used as "throwaway" inserts and are supported in special holders or shanks. The inserts have from three to eight cutting edges, and when one becomes dull, the insert is quickly indexed to a new cutting edge until all edges are worn out.

In the last few years, coated sintered carbide tools have been developed, allowing for higher cutting speeds and higher temperatures. Production rates can often be increased by 200% by switching from conventional carbides to coated sintered carbides. The coating may consist of titanium carbide, titanium nitride, or aluminum oxide.

Ceramics usually consist of pressed and sintered aluminum oxide. For light finishing cuts, cutting speeds are two to three times faster than cutting speeds for sintered carbides. Generally, they are used only for high surface finishes and close tolerances.

Diamonds are the hardest tool materials available. They are used for nonferrous machining where very high surface quality and close tolerances are required. Diamond tooling is also used to cut very abrasive materials such as composites. Diamond tools can produce mirrorlike surfaces. Since these tools are expensive, they are used only when they can be justified economically.

CBN (cubic boron nitride) is a relatively new hard cutting material that allows for high cutting speeds and temperatures. It is currently used for machining white cast iron and superalloys.

Cutting Process Conditions

This section will discuss some of the parameters of cutting operations and explain how varying them affects cutting processes. The following three parameters are commonly adjusted to optimize cutting operations (see Figure 2–11):

* Cutting speed
* Feed rate
* Depth of cut

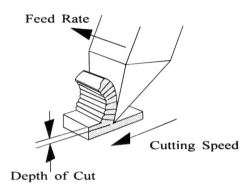

Figure 2–11. Process condition variables (mechanical mass-removal processes)

It is important to select the proper conditions for any specific cutting operation. There are usually two methods for choosing the process conditions or parameters. The most common selection method is based on economic considerations. The goal of this method is to carry out the process as inexpensively and efficiently as possible. This is done by selecting cutting speeds, feed rates, and depths of cut that will extend the life of the tool and reduce tool replacement costs. These conditions can usually be determined experimentally. A second method for choosing the process conditions is based on maximizing production. Feed rates and cutting speeds chosen for high production increase the severity of tool wear, requiring tools to be replaced more often. Another consideration when choosing process parameters is that cutting speed, feed rate, and cutting depth must be kept within the capabilities of the machine. Often, cutting speed is kept within certain limits, and feed rates and depths of cut are varied.

Cutting speed, feed rates, and depths of cut depend on the type of tool being used, the tool material, tool quality, desired tool life, process machine capability, and the work material. Typical cutting speeds for high speed steel tools are usually between 65 and 275 fpm (feet per minute). Cutting speeds for tools made of sintered carbides range from 260 to 1300 fpm. Feed rates are generally below 0.005 in./rev. Depths of cut range from 0.005 to 0.05 in. for single-point operations and from 0.002 to 0.5 in. for multipoint operations such as milling.

The power required to perform an operation can be estimated in several ways. One formula for estimating the power requirements [in kilowatts (kW)] for cutting ferrous materials is given below:

$$N = \frac{10\text{HB} \bullet f \bullet v \bullet a}{60} \qquad (2\text{–}3)$$

N = power requirement (kW)
HB = Brinell hardness number

f = feed rate (in./rev)
v = cutting speed (in./min)
a = depth of cut (in.)

The number 60 in (2–3) converts minutes to seconds. This equation is based on experiments that indicate that it takes approximately 10 times the Brinell hardness number in energy to remove one volume unit of material. Since ferrous materials are usually fairly hard, softer materials would require less power than indicated by this formula. A method for estimating the horsepower requirement for cutting both ferrous and non-ferrous materials is given below:

$$\text{hp} = V \bullet u \tag{2–4}$$

hp = horsepower
V = removal rate (in.^{3}/min)
u = unit power (hp-min/in.^{3})

Unit power, u, has been found experimentally to be approximately 0.3 for aluminum, 1.0 for soft steel, and 2.2 for hardened steel.

Another consideration in choosing cutting conditions is the desired surface quality or surface roughness. Surface finish is a function of the feed rate and the radius of the nose of the tool. For turning operations, surface roughness is determined by the arithmetic average. This value can be approximated by the following formula:

$$R_a = \frac{31f^2}{r} \tag{2–5}$$

R_a = arithmetic average (surface roughness in microinches)
f = feed rate (in./rev)
r = tool nose radius (in.)

For face milling, the maximum roughness value, R_{max}, can be approximated by the following formula:

$$R_{max} = \frac{250f^2}{D} \tag{2–6}$$

R_{max} = maximum roughness (microinches)
D = diameter of cutter (in.)
f = feed rate (in./rev)

If a certain workpiece roughness is required, the feed rate and the tool selected must be suitable.

Cutting fluids can sometimes improve cutting conditions. The unwanted effects of high friction, high cutting edge, and temperature can be reduced by introducing suitable cutting fluids into the tool–chip interface area. The purposes of cutting fluids include the following:

* Reduce friction and wear
* Cool the cutting edge
* Protect the new surface against corrosion
* Flush away chips

Work Materials (Cutting)

The work material must possess properties that permit machining to occur reasonably. The material's machinability is defined as the characteristics of the material that affect how well it can be machined by a certain process. Materials with high machinability ratings are easy to machine. Those with low machinability ratings are difficult to machine. Machinability can be rated in terms of tool wear, surface quality, cutting forces, or types of chips produced. Tool wear is frequently considered the most important factor in the machinability index. The most common workpiece materials are ferrous and nonferrous metals, but ceramics, wood, and plastics are also machined.

The machinability of a particular material is affected primarily by the material's hardness, composition, and heat treatment. A hardness range of 170–200 HB is optimal, but materials with hardness up to 300 HB can be machined with carbide cutting tools. Materials with hardness up to 600 HB can be machined with ceramic tools. Material composition is important, since, along with heat treatment, it determines the material's properties. Some alloying elements such as sulfur improve the machinability of steel. Lead, tellurium, selenium, and bismuth have similar effects.

Heat treatments also significantly influence the machinability rating of a given material. Materials that have been heat treated to achieve a coarse-grained structure generally have a better machinability rating than those with a fine-grained structure. Inclusions, hard constituents, scale, oxides, and other hard impurities that result from the steelmaking process negatively affect the machinability rating of a given workpiece.

The state of the machine tool (static or dynamic) and the rigidity of the tool and workpiece setup also greatly influence the actual cutting process.

Characteristics of the Cutting Processes

Components produced by cutting processes have the following characteristics:

* Close dimensional tolerances
* High surface quality

The following are some characteristics of the cutting process itself:

* A wide range of workpiece geometries can be produced
* A variety of materials can be processed
* Cutting processes are adaptable to numerical control and automated production techniques
* Coupled with numerical control and automated production, cutting processes produce high quality products consistently and economically

2.4 Reducing (Abrasive Machining Processes)

General Definition

Abrasion processes can be characterized by the following:

* Material removal occurs by fracture (and erosion)
* Removal rates are relatively slow
* Abrasive particles are used to remove material

The three types of abrasion processes are classified according to how the abrasives are manipulated. Listed below are the differences among each type of abrasive machining process:

1. Abrasives are bonded together in a rigid tool—this method is used for grinding
2. Abrasives are suspended in a liquid—this method is used for ultrasonic machining
3. Abrasives are directed through a jet, either by a liquid or by air—this method is used for abrasive jet machining

Workpiece shape depends on which type of abrasion process is used. If method 1 is used, workpiece shape is also determined by the shape of the tool. Workpiece shape also depends on the relative motion between the tool (or jet) and the workpiece. In industry, the first method is most commonly used.

Process Mechanics (Abrasive Machining)

In all three methods, particles remove small chips from the work material and produce good surface quality. However, in discussing process mechanics, it is simpler to consider the three abrasion methods separately.

Method 1. Rigid Tools (Grinding) Grinding "tools" have abrasive particles rigidly bonded together by clay, metal, shellac, plastics, or other suitable materials. Grinding is a very important industrial process for accurately finishing hard metals, soft metals, and nonmetals. Rigid tools can be produced in various shapes to perform special cutting operations. Figure 2–12 illustrates the abrasion principle. Surface grinding is a multipoint cutting process in which several small chips cut simultaneously. When the cutting edges (abrasive particles with undefined edge geometry) are dull, old particles are either broken out or fractured, thus producing new, sharp cutting edges. In this way, grinding wheels are self-sharpening if properly applied.

Several factors affect the process results. Depending on the particle size and type, bonding material, particle hardness, work material, and process conditions (feed rate, cutting speed, depth of cut), a wide variety of surfaces and tolerances can be produced. The bonding material for each type of abrasive and application is important; it must be selected in accordance with the work material and type of grinding process.

Method 2. Unbonded Particles (Ultrasonic Machining) In this type of abrasion, unbonded abrasives are carried in a liquid and are confined or controlled by a rigid tool in a process known as ultrasonic machining (see Figure 2–13). The slurry (mixture of particles and liquid) is fed between the workpiece and the shaping tool, which is subjected to ultrasonic vibrations, thus driving the particles against the work material at very high velocity. The "hammering" action of the abrasive particles cuts away small chips of material and produces a shape that is determined by the tool shape and the relative motion between the tool and the workpiece.

Barrel tumbling is a finishing process that also uses an abrasive slurry. The

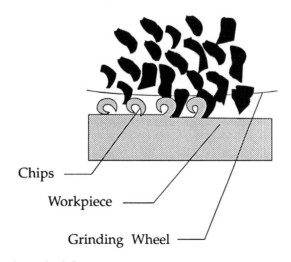

Chips ——

Workpiece ——

Grinding Wheel ——

Figure 2–12. Abrasion principle.

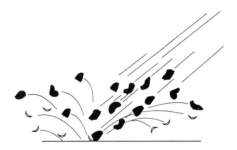

Figure 2–13. Abrasion with unbonded particles (ultrasonic machining).

workpieces are tumbled in an abrasive environment that removes a very small amount of material from the surface.

Method 3. Unbonded Particles (Jet Machining) In this process, particles are moved in an air or liquid jet. Sandblasting, which is used for surface preparation, is a similar process (Figure 2–14). Waterjet or abrasive waterjet is generally used as a cutting process.

In this type of abrasion, colliding particles contained in a high speed fluid remove small chips from the work material. The resulting geometry depends on the workpiece material and the characteristics of the jet stream (velocity, particle density, particle size, particle type, and size of the stream).

Abrasive Media Forms

The following is a list of the different forms of abrasives and the characteristics of each form:

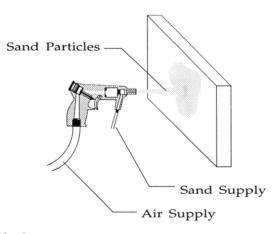

Figure 2–14. Sandblasting.

1. Bonded wheels and belts
   * Abrasive particles are bonded by a suitable bonding material
   * Grinding media come in many shapes (such as wheels or segments) and in different grades
2. Nonrigid abrasive media plus rigid shaping tools
   * Abrasive particles are carried in a liquid
   * The rigid shaping tool is subjected to ultrasonic vibrations
   * Vibrations and the feed rate determine the gap in which the particles work
3. Jets of air or liquid
   * Abrasives are carried in an air or liquid jet
   * High velocity particles impact the workpiece and remove small particles
   * Jet machining is used to produce details on glass as well as to cut plastics, honeycomb materials, sheet steel, or other materials

Abrasive Material Types

Abrasives are made of a variety of hard, brittle, crystalline materials, including silicon carbide, aluminum oxide, boron carbide, and sand. Silicon carbide is used mostly for abrasive machining of nonferrous materials, nonmetallics, and cast irons; aluminum oxide is used for both hard and soft steels; boron carbide is used for precision grinding; and sand is used for low-cost applications such as deburring and surface preparation.

Bonding Materials for Abrasives

Bonding materials hold abrasives together to form various grinding wheel shapes and fasten abrasives to flexible belts. Common bonding materials for rigid wheels include clay, rubber, and resins. Metallic bonding is a method used for fastening very hard abrasives (such as diamond or cubic boron nitride particles) to a metal grinding wheel. Grinding wheels of this type have a true roundness and can be used to achieve very close tolerances. Bonding materials for thin, flexible abrasive cutoff wheels and flexible belts include shellac and plastic bonding materials. The properties of grinding tools are determined by particle density and size, type of abrasive, type of bonding material, and bonding material hardness.

Process Conditions (Abrasive Machining)

The main conditions that may be adjusted to achieve optimal results are given below for each of the different groups of abrasion processes:

Grinding Processes Factors that affect conditions include abrasive material, bonding material grade, cutting speed, feed rate, depth of cut, and coolant. These may all be varied according to the material to be ground, the surface finish desired, machine tool rigidity, and the overall economics of the process.

Ultrasonic and Other Slurry Processes Factors that affect conditions include particle size and type, type of liquid carrier, ultrasonic frequency, and amplitude. These may be varied to suit the type of material being machined, to achieve desired removal rates, and to maintain process economics.

Jet Machining Factors that affect conditions include particle type and size, size of jet orifice, fluid material, and jet velocity.

Workpiece Materials (Abrasive Machining).

Generally, the following properties of workpiece materials are important:

* Strength
* Ductility
* Hardness

Grinding is used on hardened steels, carbides, cast iron, aluminum, brass, copper, and bronze. Ultrasonic machining is used on carbides, ceramics, glass, gemstones, or synthetic crystals. Jet machining is used for fragile materials, for frosting glass and in the case of water-jet cutting for virtually all types of materials.

Characteristics of Abrasive Machining Processes

Abrasive machining processes are important to modern manufacturing. Generally, these processes offer high accuracy (narrow tolerances) and relatively fine surface finishes.

2.5 Separating Processes

General Definition

An important subfamily of the mechanical material-removal processes includes the separation, or shearing, processes. Separation (shearing) processes have the following characteristics:

* Shearing is a mechanical process
* Shearing is a controlled fracture that occurs in a well-defined, narrow shearing zone (see Figure 2–15)
* The shearing zone is defined by an upper tool and a lower tool

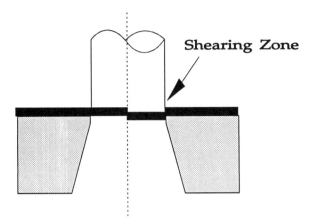

Figure 2–15. Creation of a shearing zone.

* The resulting workpiece shape is determined by the shapes and the relative motions of the upper and lower shearing tools

The main element of shearing is the creation of a narrow shearing zone in which fracture occurs.

Process Mechanics

In shearing processes, the upper and lower tools determine the shearing zone, which normally has a width that is 1 to 10% of the thickness of the work material.

If supporting elements are used, a more narrow zone (1%) can be established, allowing for better product quality. This process is called fine blanking, and it is more expensive than conventional blanking because it requires more complicated tooling and a more expensive press.

Figure 2–16 illustrates a bit more detailed picture of shearing mechanics. When the punch starts to move into the material, it penetrates it along the cutting edge, and, at the same time, the workpiece deforms along the edges of the die. When the penetration takes place, fractures are initiated along the tool edges. With increasing punch penetration, the fractures dominate, and the two material parts are separated. The major parameters that determine the quality of the shearing process are the following:

* Thickness of the material
* Material type and hardness
* Clearance between cutting edges
* Quality (sharpness) of cutting edges

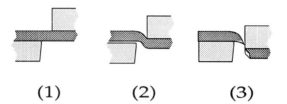

(1)   (2)   (3)

Figure 2–16. Shearing mechanics (with exaggerated clearance).

* Support of material
* Diameter of hole or blank in relation to material thickness

Tool clearance is normally between 1 and 10% of the material thickness depending on the remaining parameters. If strong support exists, the clearance can be reduced, resulting in better quality.

Tool Types

Even though the tool geometries of these processes are different, they all utilize the same basic shearing mechanism.

Tool Materials

In the shearing processes, heavy loads are present at the cutting edges, and extensive wear takes place. The following factors affect tool wear:

* Properties and dimensions of the workpiece material
* Production quantity
* Required quality

The following tool materials may be selected:

* Plain carbon steels
* Tool steels
* High speed steels
* Sintered carbides

Normally, tool cutting edges must be sufficiently hard to be adequately wear-resistant.

Process Conditions

The main parameters in the shearing process are as follows:

* Clearance (1–10%)
* Speed (impact)

* Temperature
* Work material (type, properties, dimensions)
* Conditions of cutting edges
* Support of work material

The approximate necessary force may be calculated from the following formula:

$$P = L \bullet t \bullet S \qquad (2\text{-}7)$$

P = approximate force required
L = length of cut
t = material thickness
S = shear strength

If the shear strength, S, is unknown, the ultimate tensile strength σ_{ult} can be used, as shown in the following formula:

$$P = 0.8L \bullet t \bullet \sigma_{ult} \qquad (2\text{-}8)$$

when it is assumed that $S = 0.8\sigma_{ult}$.

The preceding formulas produce acceptable results. If hot shearing is used, the S or σ_{ult} must be reduced correspondingly.

Workpiece Materials

The following workpiece material properties are important:

* Hardness/ductility—must show a ductile fracture (HB < 200)
* Strength
* Composition and homogeneity

Typical workpiece materials include the following:

* Ferrous and nonferrous metals
* Plastics
* Paper, cardboard, etc.

In the metalworking industry, squaring, blanking and piercing are used extensively.

Characteristics of Separating Processes

If no workpiece support is present, the tolerances obtained will vary according to the parameters used. Typical tolerances lie within ±0.01 in. to ±0.060 in. In fine blanking (which uses a special support), tolerances of ±0.0005 in. can be achieved. The surface finish in conventional shearing is rather poor (100–200 microinches), but in fine blanking, surface finish is good (30–50 microinches).

Thermal Mass-Reducing Processes

3.1 Introduction

The thermal mass-reducing process family is listed in the taxonomy (Chapter 1) under mass-reducing processes. Mass-reducing processes are shaping processes. Thermal mass-reducing processes are used extensively in industry because they are capable of cutting very hard materials with minimal mechanical force.

In general, thermal mass-reducing processes have the following characteristics:

* The desired shape is obtained by removing material from a solid workpiece material
* Material removal takes place thermally by local melting, evaporation, or combustion
* Thermal energy is created either chemically or electrically
* Creation of the desired shape is based on relative motions between the heating source (tooling) and the workpiece

3.2 Basic Processing Energy

The processing energy in this family is thermal energy. Thermal mass-reducing processes use local melting, local evaporation, or local combustion to remove material and produce the desired geometry.

Different methods of applying thermal energy have been developed. There are two fundamental types of thermal mass reduction used in industry:

* Type I.   Mass reduction without using a shaping tool
* Type II.   Mass reduction using a shaping tool

Type I processes (Figure 3–1a) are characterized by direct interaction between the energy source (e.g., flame or laser beam) and the workpiece; these processes include torch cutting and high energy beam machining processes. Type II processes (Figure 3–1b) use a shaping tool with the thermal energy source to remove material and form the workpiece; these processes include electrical discharge machining (EDM) and wire-electrical discharge machining (wire-EDM).

3.3 Process Mechanics

Type I Processes

Typical energy sources for Type I processes include flames, beams (laser and electron), and arcs (cutting electrodes). The energy source is described by the following parameters:

* Intensity (watts/in.^{2}) and distribution
* Temperature of the heat source
* Dimensions of the heat source

The energy source interacts with the workpiece material. The result of this interaction depends on the characteristics or properties of the heat source and the properties of the workpiece material. The most important workpiece material properties include the following:

* Heat conductivity
* Melting point
* Chemical composition
* Combustion properties
* Light absorption and reflection properties

Normally, combustion of the workpiece material in the cutting zone takes

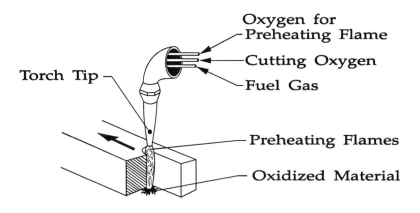

(a)

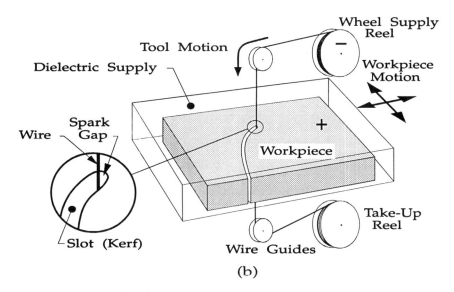

(b)

Figure 3–1. Two examples of thermal mass reduction: (a) Gas flame cutting (Type I) and (b) wire-EDM cutting (Type II).

place when oxygen reacts with the heat source. This combustion provides an important energy input. A steady-state condition is achieved in the following situation:

Energy from heating source + energy from combustion
= total energy used for heating, melting, and evaporation + energy losses

The molten material is blown out as drops, slag, burrs, and so on, due to pressure from the cutting gas and the energy source (Figure 3–2).

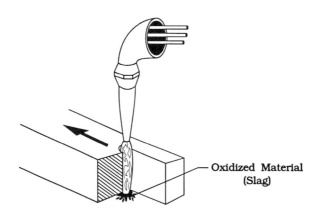

Figure 3–2. Material removal in gas flame cutting.

The material removal rate, V (in.^{3}/sec), is given by (3–1).

$$V = \frac{volume}{time} = feed \times width\ of\ cut \times thickness \qquad (3–1)$$

The approximate energy required for melting can be computed using (3–2).

$$V_o C_p T \qquad (3–2)$$

where V_o is the workpiece material density, C_p is the heat capacity, and T is the melting temperature. As mentioned, oxygen is mixed with the fuel gas to ensure good combustion for cutting. Some materials, such as stainless steel, form stable oxides that have poor cutting qualities and require mechanical deburring afterwards. Substituting part of the oxygen with an inactive (shielding) gas can help prevent formation of oxides and solve the cutting problem.

As mentioned before, the energy sources are flames created by gas torches (combustion of oxygen and acetylene) or plasma torches (hydrogen or air forms a plasma that creates much higher temperatures when heated than a gas torch).

Arc cutting is a special process in which an arc strikes between an electrode and a workpiece and creates the necessary heat to melt a certain amount of material (Figure 3–3). A shielding gas is used to prevent oxides from forming on the surface of the workpiece.

In high energy beam machining, laser beams and electron beams are used. Laser beams are light beams with a wavelength of 4.17×10^{-4} in. for CO_2 and 4.17×10^{-5} in. for neodymium-yttrium aluminum garnet (ND-YAG). Laser beams are absorbed or reflected by the workpiece material, depending on the specific workpiece material properties and temperature. Laser beams are very small and have high energy densities. Electron beams require a vacuumized atmosphere or the

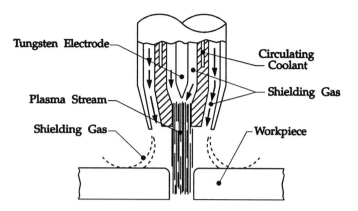

Figure 3–3. Plasma arc cutting.

electrons will lose their energy due to impact with air molecules. Electron beams may be used for cutting, but they are mainly used for welding.

The energy supply in the beam, the workpiece material properties, shape, and dimensions, and the relative speed of motion between the energy supply and the workpiece are the major process parameters to consider in high energy beam machining. These parameters depend on each other.

Type II Processes

In Type II thermal mass-reducing processes such as cavity-type electrical discharge machining, a rigid shaping tool is used to remove material. As the material is removed, the workpiece takes the shape of the rigid tool. The thermal energy is supplied, or developed, between this tool and the workpiece material (Figure 3–4).

The tool in a normal electron discharge machining (EDM) process forms the shape cut into the workpiece. In a simple kinematic pattern (a pattern with one translation), the properties in the workpiece surface are developed. Material is removed when sparks are generated between the tool and the workpiece. Sparks, frequency, and current are controlled at the energy source. To control the sparks, a dielectric fluid is introduced between the tool and workpiece. As the tool approaches the work material, sparks are generated one at a time where the distance is the shortest between the work material and the tool. Each spark leaves a small crater in the workpiece so that the final surface is a *matte* finish. If a continuously moving wire instead of a rigid shaped tool is used, the process is called wire-EDM (Figure 3–5). Wire-EDM is especially well-suited for cutting applications (such as blanking dies and net shape components).

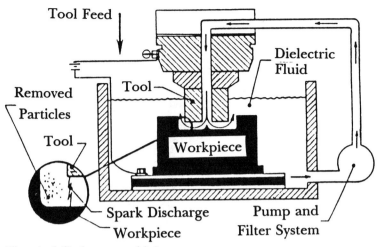

Figure 3–4. Electrical discharge machining.

3.4 Types of Tools and Tool Materials

Type I thermal processes do not require a rigid tool, whereas Type II processes do. EDM requires a three-dimensional tool whereas wire-EDM only requires the wire. To maximize process performance in EDM, the tool material should have the following properties:

* Good electrical conductivity
* Good thermal conductivity
* Relatively high melting point
* Good machinability (formability)

Typical tool materials are graphite, brass, and copper. Depending on the application, mineral oil, kerosene, or distilled water is used as the dielectric.

3.5 Process Conditions

For Type I thermal cutting processes (without rigid shaping tooling), the following parameters are important:

* Energy source (energy density, distribution, dimensions, focusing)
* Atmosphere (air, oxygen, mixtures)
* Cutting speed (NC/CNC table speed)
* Workpiece material

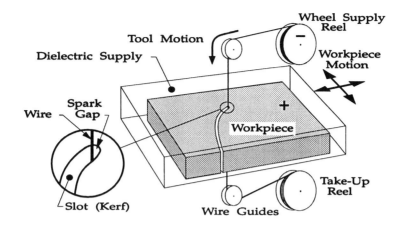

Figure 3–5. Wire electrical discharge machining (wire-EDM).

Energy Source

Energy density, or distribution of the energy source, is very important for the quality of cut. For example, the gaussian (cylindrical) distribution is acceptable for laser beams, but a square distribution may be even better. The energy level can be varied within certain limits, depending on the capability of the power supply. Different sizes of torches and lasers exist. Each torch or laser design has certain characteristics that must be studied carefully. The dimensions of the energy source are critical since they must be adapted to the thickness of the material to be cut.

Atmosphere

When the heating source interacts with the workpiece, the atmosphere can play an important role. If pure oxygen is used instead of air, the cutting quality improves considerably. In some materials with high chromium content, oxygen will create an oxide (chromium oxide), and the components cannot be welded without mechanical edge preparation.

Cutting Speed

Cutting speed is a dependent parameter. After the energy level is set and the dielectric is selected for the workpiece material, the cutting speed is then adjusted to maximize the quality of the cut. For NC/CNC cutting systems, the problem is easy to handle.

Workpiece Materials

Materials suitable for thermal mass reducing or cutting include ferrous and non-ferrous metals as well as nonmetallic materials (for laser and plasma cutting). The higher the thermal conductivity and the higher the reflective properties of a workpiece material, the more difficult the material is to cut.

For Type II mass-reducing processes, the main parameters include the following:

* Current (current density)
* Discharge frequency
* Spark gap
* Feed

The dielectric fluid and the tool (electrode) material must be selected according to the rules mentioned in the list at the beginning of section 3.5.

The current and frequency determine spark size, and the spark size determines the crater dimensions and the surface roughness. Normally, a rough cut is made first, followed by a fine cut. For a certain setting of current and frequency, the speed is selected so that a reasonable gap exists between the electrode and the workpiece. Feeds that are set too high will lead to short circuits. For wire-EDM, the same conditions are valid, except that distilled water is often used as the dielectric fluid.

3.6 Workpiece Material Properties

The important properties for workpiece materials in Type I processes are as follows:

* Thermal conductivity (reasonably low)
* Melting point (reasonably low)
* Composition
* Reflectivity (as low as possible)

For flame cutting, steel is the most common workpiece material. For laser cutting, steel, nonferrous materials (except copper, silver, gold, and their alloys because they have very high conductivity and reflectivity), and a broad range of nonmetallic materials are commonly used workpiece materials.

For Type II processes, the main requirement for a workpiece material is that it be electrically conductive. Typical materials include steels, cast iron, aluminum, tool steels, and cemented carbides.

3.7 Characteristics of Components Produced by Thermal Mass-Reducing Processes

Based on the general description of thermal mass reducing, expect the following characteristics for Type I and Type II processes:

Type I Processes (without shaping tools)
Flame/Plasma Cutting
* Poor surface finish
* Coarse tolerances
* Large heat affected zone
* Heavy burrs

Laser Cutting
* Relatively good surface finish
* Fine tolerances
* Narrow heat affected zone
* Minimal burrs

Type II Processes (with shaping tools)
* Good surface finish
* Fine tolerances
* Narrow heat affected zone with thermal microcracks
* No burrs

Shaping tools produce better quality surfaces than do nonshaping tools.

Chemical Reducing Processes

4.1 Introduction

Chemical reducing processes are listed under shaping in the mass-reducing family of the taxonomy chart (Chapter 1). These processes generally have the following characteristics:

* The desired shape is obtained by removing material from a solid work material
* Material removal occurs chemically by dissolution of the material
* Chemical dissolution is created by pure chemical etching or electrochemically

4.2 Types of Chemical Reducing Processes

There are two types of chemical reducing processes (Figure 4–1):

1. Pure chemical dissolution with an etching solution
2. Electrochemical dissolution using an electric charge coupled with an electrolyte

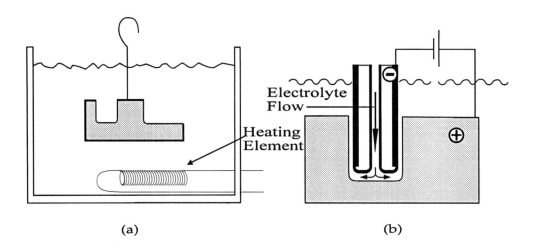

(a)   (b)

Figure 4–1. Two principles of chemical mass-reducing processes: (a) Etching and (b) electrochemical machining.

4.3 Process Mechanics

In both of these processes, material is removed by controlled chemical dissolution. In etching, the parts to be removed are chemically attacked. The part of the workpiece that must not be affected is protected by a masking material. With etching, no tool is used (Figure 4–1a). The second type of chemical reducing process, electrochemical machining, is based on chemical dissolution that is driven electrolytically. A shaped tool is fed down into the work material while the electrolyte is being flushed between the electrode and the work material (Figure 4–1b). A DC current drives the dissolution. The resulting cavity is the same shape as the tool.

Etching can be described by the following reactions:

* Anode reaction: $M \Rightarrow M^{n+} + ne^-$
* Cathode reaction: $Z^{n+} + ne^- \Rightarrow Z$
* Total reaction: $M + Z^{n+} \Rightarrow M^{n+} + Z$

$$M = \text{metal}$$
$$Z^{n+} = \text{oxidizing ion in the solvent}$$
$$n = \text{number of electrons removed}$$
$$e^- = \text{electrons}$$

Essentially, a solution is chosen that will remove electrons from the metal in order to reduce the oxidizing ions in the solvent to a non-ion form. The selection

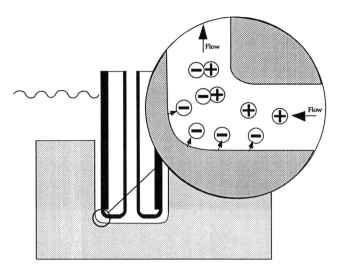

Figure 4–2. The principle of electrochemical machining.

of the chemical solution depends on the material to be etched. When the electrons are removed from the metal and combine with the oxidizing ion in the solvent, metal ions go into solution; in other words, the metal is dissolved. The main parameters in this process are solution concentration, bath temperature, and immersion time.

On the molecular level, electrochemical machining is similar to simple chemical machining, except that the etching solution used is not strong enough to remove material by itself. This problem is overcome by passing an electric current through the chemical solution to increase its chemical power. Electrons are removed from metal, causing metal ions to go into solution. Depending on the materials and the chemicals, different reactions can occur. In the most common reaction, metal ions combine with ions in the solution and precipitate out as a sludge. When using stronger etching solutions, the metal ions either stay in solution or tend to plate out on the tool. The parameters in this process are the applied voltage, electrolyte concentration, electrolyte pressure, bath temperature, and electrolyte flow rate (Figure 4–2).

4.4 Types of Tools and Tool Materials

Rigid tooling is not used for etching. Etching involves the following components (Figure 4–1):

* Etching solution
* Masking material

The etching solution can be either alkaline or acidic depending on the work material to be etched. The solution may also be heated to aid the process. The masking material covers and protects the areas of the workpiece that are not to be etched. Masking materials include adhesive tape, copper electroplate coating, paint, plastic, and photosensitive material.

For electrochemical processes, the tooling system consists of the following components (Figure 4–2):

* Shaped electrode
* Electrolyte

The electrode must be easy to machine, corrosion resistant, and electrically conductive. Also, it must be homogeneous and can be composed of copper, brass, stainless steel, or graphite. Commonly used electrolytes are solutions of sodium nitrate and sodium chloride (salt water).

4.5 Process Conditions

For etching, the process parameters are the following:

* Type and concentration of the etching solution
* Temperature of chemical bath
* Method of circulating the etching solution (includes spraying the etching solution)
* Time that the workpiece is left in the chemical bath
* Workpiece material

For each specific workpiece material, these parameters must be controlled for best results. The masking material and the above parameters determine the accuracy of the workpiece's final shape. For electrochemical machining, the process parameters are the following:

* Voltage and current density (amps/in.^{2})
* Type and concentration of electrolyte
* Temperature of the tool, workpiece, and electrolyte
* Shape of the electrode
* Feed rate

It takes a great deal of experience to design an electrode that actually produces the desired geometry. Good results depend on careful parameter control.

4.6 Workpiece Materials

For best results in chemical processing, the following workpiece material properties are required:

* Uniform grain size
* Homogeneous structure (free from cavities and impurities)
* Electrically conductive (a requirement for materials that are processed electrochemically)

Workpiece materials include the following:

* Alloyed steel
* Aluminum
* Magnesium

One important characteristic of chemical machining processes is that workpiece hardness or toughness is not a factor. Workpieces may be processed in their hardened state. Another characteristic of these processes is that no surface stresses are produced.

4.7 Characteristics of Workpieces Produced by Chemical Reducing Processes

For both chemical and electrochemical processes, a reasonably close tolerance level can be achieved, as well as good surface finish. Tolerances and surface finish depend on the grain size, porosity, and purity of the workpiece material. Parameters such as temperature, solution concentration, immersion time, and circulation of the chemical influence the process results.

Consolidation Processes

5.1 Introduction

Consolidation processes are mass-conserving processes; they are found in the taxonomy chart in the shaping branch (Chapter 1). The five categories of consolidation processes are as follows:

* Compacting
* Casting
* Molding
* Laminating
* Depositing

These process categories can be subdivided into more specific categories, depending on the type of workpiece material and mold material used. This chapter will discuss some of the processes in each category, with the exception of deposition processes, which are listed as electroforming in the taxonomy chart.

There are four basic steps in consolidation processes:

1. The workpiece material is conditioned and prepared for shaping
2. Workpiece shaping is accomplished by a process such as compacting, molding, or laminating
3. The shape of the workpiece is stabilized

4. Finishing operations and/or postsintering treatments are carried out

As mentioned, consolidation processes can be grouped according to the workpiece material used. For example, casting and molding processes involve a liquid workpiece material. Compacting processes involve a granular workpiece material.

5.2 Compacting Processes with Granular Workpiece Materials

Compacting processes use metal and carbide powders (grains) as the raw material to produce coherent metal and carbide parts. The manufacture of products from the granular or particle state covers a broad spectrum of materials and components, such as metal powder components, sand molds, ceramics, concrete, and tableted materials.

Granular workpiece materials may be used for the following reasons:

1. The material is only available in the granular state
2. The desired properties (porosity, combination of materials, etc.) can be achieved only in this way
3. Manufacture of the product is cheaper this way
4. Small components may be difficult to produce otherwise

In this book, only powder metallurgy (production of components from metal powders and carbides) will be discussed. The production of components from metal powders has increased in recent years due to the capability to produce a near-finished shape in one operation with only limited postworking, the ability to produce intricate shapes in a single operation, a high material utilization, and a low unit price.

Basic Processes

Shaping is accomplished by two basic processes through mechanical processing energy: plastic deformation and the flow of particles. When compacting metal powder at a pressure above the particles' yield strength, plastic deformation will appear in the contact zones, enabling "cold welding" of the particles to occur. To fill the die cavity properly, the powder must flow uniformly, freely, and rapidly into the cavity. The flow rate of any powder is greatly influenced by the size and shape of the powder particles, the distribution of the different particle sizes, and the freedom of the particles from moisture. Through the flow characteristics of the powder and the compaction process, which is caused by motion in both the upper and the lower punches in a double-acting press, a distribution with a uniform density can be obtained (Figure 5–1a). It is important to note that flow that is perpendicular to the direction of compaction is very difficult and should not be considered.

After compaction, the component is in its green (unfinished) stage, with sufficient strength to be handled manually or by automated machinery. The component is then stabilized through a heat treatment process called *sintering*. Sintering can be carried out in the solid phase or in the combined solid/liquid phase, where one of the ingredients has a lower melting point than the others and is in the liquid phase and the other ingredients are in the solid phase. Sintering with a liquid phase will create a rather dramatic shrinkage of dimensions, but still produce remarkably accurate parts.

Process Mechanics

There are several methods of compacting metal and carbide powders:

* Conventional compaction (Figure 5–1a)
* Isostatic compaction (Figure 5–1b)
* Extrusion compaction
* Rolling compaction

A binder is a cementing medium that is added to the powder to increase the green strength (caused by the interlocking and cold welding of the particles during compaction) or the sintering strength of the component. Methods of compacting carbides included in a binder are as follows:

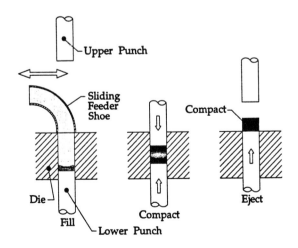

Figure 5–1a. Conventional powder compaction.

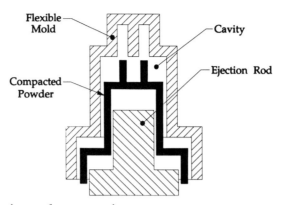

Figure 5–1b. Isostatic powder compaction.

* Compaction
* Slip casting

Slip casting of carbides, in which the carbides are suspended in a liquid and then casted, will not be discussed.

The stages associated with powder compaction are the following:

* Conditioning and preparation for compacting and sintering
* Flow and plastic deformation (compaction)
* Sintering
* Postsintering treatments

Conditioning and preparation requires mixing the necessary ingredients, which include powders, alloying elements, carbon, and lubricants. Mixing must be done within rather short time limits; otherwise, plastic deformation or separation of particles may result. The conditioning phase must also be performed carefully to ensure a high quality product.

The flow process aims at maintaining a uniform powder distribution in the die cavity so that the density distribution in the compacted part is uniform. Normally, flow is controlled in the die design through the application of split lower punches that travel in such a way that the compression ratio (height of loose powder divided by height of powder in the green compact) is the same for all levels. When compaction occurs, all areas of different thicknesses in the component must be compacted to the same density so that a uniform density distribution is obtained (Figure 5–2).

Friction between powder and die walls, between powder and punch surfaces, and between powders normally creates a nonuniform density distribution. There-

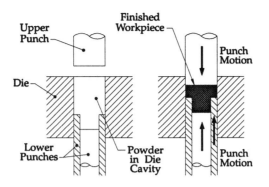

Figure 5-2. Two-sided powder compaction with floating or guided container and split lower punches.

fore, in conventional compaction, the component is compacted by punches pressing from both sides (double-action pressing), and the die is either floating or controlled by kinematics. The powder grains are plastically deformed and cold welded during compaction. The pressure required to create a certain density by plastic deformation is shown in Figure 5-3 for iron-based powders. Relative density is often used to measure compaction:

$$Relative\ density = \frac{obtained\ density}{density\ of\ solid\ material}$$

Normally, relative densities are 70–95% of the solid material density.

The above discussion is based on compaction with rigid tooling. *Hydrostatic*

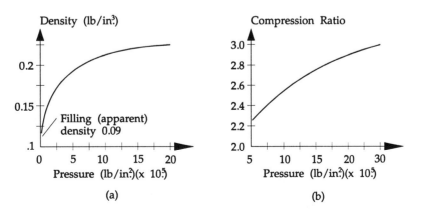

Figure 5-3. Density as a function of compaction pressure for iron powder.

compaction, in which a high liquid pressure is applied uniformly along the surface of a part, can eliminate friction between the tool and the powder and can also eliminate complicated, expensive tooling. In hydrostatic compaction processes such as isostatic compaction (Figure 5–1b), powder fills a deformable or flexible plastic or metallic liner and is then compacted by fluid pressure in a high pressure chamber at either low or high temperatures (cold and hot isostatic compaction). The compacted component maintains the original shape of the liner but is smaller.

There are various ways of applying the flow and plastic deformation principles in compacting components. Other examples of compaction processes include powder extrusion and powder rolling.

When cemented carbides are produced, the process phases are the same as those involving other carbides. For example, carbide particles are mixed with cobalt powder and compacted. The cobalt, which is actually the carrier of the carbides, forms a coherent component. During sintering, the cobalt melts, and about 20–30% shrinkage occurs.

In general, sintering is a very important step in the production of powder metallurgy products. Sintering mechanisms create the most strength in the finished compacted component. If carried out in a protective atmosphere to prevent oxidation, sintering requires a rather high temperature. Heat is an important consideration in surface and volume diffusion processes. The following processes are heat-dependent:

* Surface diffusion
* Volume diffusion
* Plastic deformation
* Evaporation or conduction

In solid-state sintering (sintering at a temperature below the lowest melting point of all the ingredients), the accompanying shrinkage is about 0.1–0.3% by volume. For different materials, different temperature and time values are optimal.

When the sintering temperature is above the melting point of one of the major constituents, sintering occurs in the liquid phase. A shrinkage of 40–60% results, giving lower accuracy than solid-state sintering. Cemented carbides (such as tool bits, inserts, and bushing) are produced in this way.

Postsintering operations may include various combinations of sizing, coining, presintering, sintering, impregnation, infiltration, hardening, surface coating, and machining. Postsintering processes normally aim at increasing strength properties or reducing porosity or both. Compaction considerations can be summarized as follows:

* The higher the density, the better the properties produced
* The higher the density, the higher the required die load
* Two-sided compaction produces a more uniform density distribution
* Compaction ratios range between 2.5 and 3.5

Tool or Die Materials

Depending on the specific process variants, different tool materials are used. Conventional powder compaction uses the following tool materials:

* Hardened tool steels (for parts in contact with powder)
* HSS (high speed steels) or cemented carbides are often used for containers and punches. The container and the punch must be polished.

Isostatic powder compaction uses the following tool materials:

* Low alloy steel casing
* Plastic and rubber

In general, handling powder under high pressure and sliding conditions requires extremely hard and wear-resistant tool and die materials.

Process Conditions

The main parameters for powder-based components are as follows:

* Type and properties of powder
* Compacting pressure (depends on relative density)
* Density distribution
* Sintering parameters (time, temperature, atmosphere)
* Type of die system

To obtain a reasonable relative density, the applied pressure (P) is normally between 1.2σ and 2.2σ, where σ is the average yield strength of the solid material.

Workpiece Materials

The following powder properties influence the process and the resulting workpiece properties:

* Type of powder
* Particle size and shape
* Compressibility of particles
* Ductility of particles (some ductility required)
* The oxides must be reducible in the sintering atmosphere

Typical powders include the following:

* Iron
* Low to medium alloyed steel
* Brass
* Copper

Typical applications include the following:

* Bearings
* Machine components (gears, levers, bushings, etc.)
* Filters

Characteristics of Components Produced from Granular Workpiece Materials

Components produced in powder metallurgy may have many excellent properties. Powder components produced with σ_{uts} of 100,000 lb/in.^{2} are not considered exceptional in today's industry. Obtainable tolerances will, depending on the actual procedure, lie within axial tolerances of 0.010–0.001 in./in. and radial tolerances of 0.001–0.005 in./in. To obtain the best tolerances, some postsintering operations are necessary. Normally, one can expect good accuracy, much geometry formed per operation, high productivity, low unit cost, and reasonable mechanical properties from powder metallurgy.

5.3 Casting/Molding Processes with Liquid Workpiece Materials

The following characteristics exist for casting and molding processes:

* The processes are mass conserving
* Liquid or plastic work materials are used
* The process can be structured in three phases as follows:
  1. Melting or plastification of the material from heating
  2. Shaping based on a mechanical process
  3. Stabilization based on cooling of the finished part
* Shaping or surface creation is established through the following:
  1. A given shape, size, and surface finish of the tool or the die
  2. A corresponding relative motion between the work material and the die

This process family includes two major industrial process groups: casting (metallic work materials) and molding (plastic work materials).

Types of Basic Processes

Creating a component from a liquid state requires three types of basic processes (Table 5–1). First, the material is melted, then it is forced into a die cavity, and, finally, the shape is stabilized by solidification, or hardening.

Casting

In casting, metallic components are shaped in the liquid state. This subfamily can be further characterized by the following stages:

1. Design and production of the mold or die system
2. Melting of solid work material
3. Introduction of molten metal into the mold or die cavity
4. Solidification of shaped molten metal
5. Removal of solidified component from mold or die

The shape is determined either by the cavity shape alone or by contact with the mold or die and relative motions between the mold or die and the workpiece material (Figure 5–4).

Process Mechanics and Process Conditions

Typical stages in the casting process are the following:

* Mold or die system design, production, and materials
* Melting
* Introduction or pouring of molten metal into mold or die
* Solidification
* Removal or extraction
* Cleaning and inspection

Table 5–1. Stages and basic processes in producing shaping components from liquid state of material casting

Stage	Nature of basic process
Melting or plastification	Thermal
Shaping (the liquid is introduced in the die system)	Mechanical
Stabilization (solidification)	Thermal

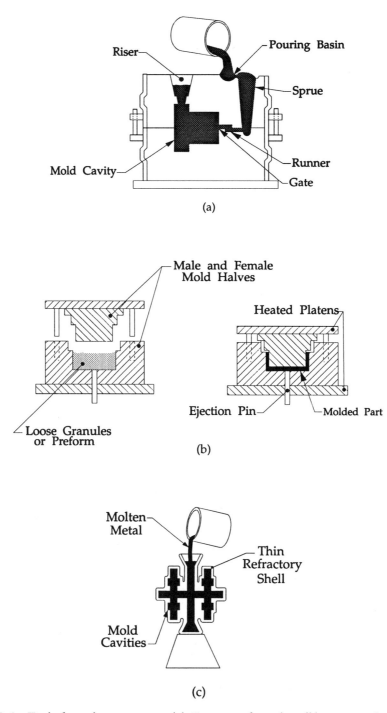

Figure 5–4. Typical casting processes: (a) Green sand casting, (b) compression casting, and (c) investment casting.

Several of these steps will be described briefly.

Mold or Die Design, Production, and Materials In general, the mold or die system must be able to do the following:

* Withstand pouring or injection pressure of the liquid metal (erosion)
* Withstand high temperatures
* Conduct heat away
* Accept or allow air or gas to escape

The design of a mold or die depends mainly on the following three factors:

Type of molds:
* Permanent (nonexpandable)
* Nonpermanent (expandable—used once)

Category of pattern:
* Permanent
* Nonpermanent

Pouring principle:
* High pressure
* Low pressure
* Gravity

Table 5–2 lists four basic groups of casting processes used. These groups differ based on the type of mold, mold material, pouring principle, pattern material, and process name.

The mold or die material chosen determines what pouring pressures can be used. A high temperature combined with a high pouring pressure places very strict requirements on these materials. High quality tool steels must be used to obtain acceptable productions without frequent replacement or repair of the die.

Melting To fulfill the requirements of material properties and to produce a sound casting, the molten metal must have the right composition and limited metallic and nonmetallic contaminations, including gases. Depending on the raw materials, a refining process or a correction of composition of the melt might be necessary. Molten metal can dissolve greater amounts of gaseous contaminations than solid metal can. During pouring, gas is precipitated, causing porosity in the casting. Many different melting processes and furnaces are available. Each produces molten metal with different "hereditary" properties (such as machinability and cooling properties). These properties influence the

Table 5–2. Types of molds/dies, mold materials, patterns, and pouring principles classified in four basic process groups

Grouping	Type of mold	Mold material	Pouring principle	Pattern material	Process name
Sand casting	Nonpermanent (single-purpose)	Sand (green)	Gravity	Wood, metal, plastics	Green sand, dry sand, core sand casting
Permanent (metallic) mold casting	Permanent	Alloy steels	High pressure	—	Die casting
		Graphite, steel, cast iron	Low pressure	—	Low pressure (permanent mold casting)
		Cast iron, steel	Gravity		Nonpressure gravity permanent mold casting
Precision casting	Nonpermanent (single-purpose)	Nonmetallic (sand, plaster, ceramics, etc.)	Gravity (low pressure)	Metal	Shell mold casting
Investment or precision casting				Wax, plastic, (rubber, metal)	Plaster mold casting
Investment or precision casting				Wax, plastic, (rubber, metal)	Ceramic shell mold casting
Investment or precision casting				Wax, plastic, (rubber, metal)	"Lost wax" casting (investment casting)
Centrifugal casting	Nonpermanent/ permanent	Nonmetallic/ metallic	Centrifugal forces	—	Centrifugal casting

selection of the melting process, as do size, shape, and composition of raw materials.

Pouring or Injection of Molten Metal into Mold or Die Producing a casting without defects depends on many factors. One important factor is the way in which the metal is poured into the mold or die. The pouring process involves the pouring pressure and the gating system (channels for delivering the metal into the mold or die). Pouring pressure can be classified as *high* (0.3–2.2 ksi), *low* (0.01–0.04 ksi), or *gravity*. High pressure injection die casting requires strong permanent dies. The higher the pressure, the thinner the sections and the finer the details that can be produced. Normally, low pouring pressures require permanent molds, but graphite molds may also be used.

The purpose of the gating system is to allow the liquid metal to enter the mold or die cavity at a certain rate and temperature. A poorly designed gating system may cause excessive heat loss, a high pouring temperature (too high a pouring temperature will result in poor grain structure, porosity, etc.), turbulence in the fluid stream, entrapment of gas, slag, dross, and heavy erosion. The actual gating system primarily depends on molding method, mold material, molten metal, geometry, pouring, and injection pressure. In general, from the pouring basin or injection system, the metal must flow through constantly decreasing cross sections to about 20% total reduction. Important requirements for a well-engineered gating system include the following:

* Prevent slag and oxides from littering the mold cavity
* Prevent air or gas inclusion and erosion of molds and cores
* Decrease requirements for high pouring temperatures
* Lead the liquid metal into the mold at the right place and at the correct rate; this results in castings with minimum shrinkage voids and distortion
* Minimize the amount of metal used in the gating system

When the metal is poured into the die or mold, contraction of the casting begins as it cools off. For steel, during cooling, contraction from pouring temperature to solidification temperature is 1–2%; contraction during solidification is 2.5–3%; and contraction during cooling from solidification temperature to room temperature is approximately 7.0%. Solidification shrinkage for aluminum is 6–7%.

Solidification Mold design greatly influences the solidification pattern. Solidification in sand and shell molds is relatively slow. A directional solidification, in which solidification starts at the lowest part of the gating and continues up through the mold, is desirable. The possibilities of obtaining a directional solidification depend on the design of the component. In permanent molds, solidification starts even before the die is completely filled and terminates shortly thereafter. This means that short, thin sections may close and prevent further feeding to other sections. In die casting, feeding is caused by high pressure supplied to the gating system in order to compensate for shrinkage. Modifications of the component may be necessary to obtain a sound component.

Removal or Extraction When the work material is solidified, nonpermanent molds are broken and the casting is cleaned. If permanent molds are used, the gating is blocked off. The molds are then opened and the component is ejected, usually automatically.

Workpiece Materials

Generally, materials can be cast when they have a low melting range [below 1800°C (3600°F)] and good castability. The most commonly used materials (alloys) are listed below, along with their pouring temperatures:

* Steel   1550–1700°C (2800–3100°F)
* Cast iron   1200–1450°C (2200–2650°F)
* Aluminum   600– 880°C (1100–1600°F)
* Copper   900–1100°C (1650–2000°F)
* Zinc   450– 550°C (850–1025°F)

In general, cast components have good mechanical properties and very complex geometries; they are also easy to machine. In recent years, the applications of precision casting processes have increased. This has also led to higher accuracy in sand casting.

Characteristics of Cast Components

In general, very complex shapes or geometries can be produced in casting. Precision casting (permanent mold casting) results in a high level of accuracy, with minimal or no postmachining required. Metals such as manganese, magnesium, zinc, aluminum, and some copper alloys can be die cast with good results. The higher the melting point, the higher the requirements of the mold, and the shorter the tool life. Cast iron and steel are often used in sand casting. The technology for casting has been extensively developed and can be automated.

Molding

In the molding subfamily, plastic components are shaped in a liquid or a plastificated state. This subfamily can be further distinguished by the following stages:

1. Design and production of the mold or die system
2. Plastification of the plastic material
3. Introduction or injection of the plastificated material into the die or mold
4. Solidification or hardening of the shaped material
5. Removal of solidified component from the mold or die

In general, the shape of the component is determined by the mold or die cavity alone (Figure 5–5a) or, in certain applications, by the contour of the mold or die and the relative motions that occur between the mold or die and the workpiece material as the process is performed (Figure 5–5b).

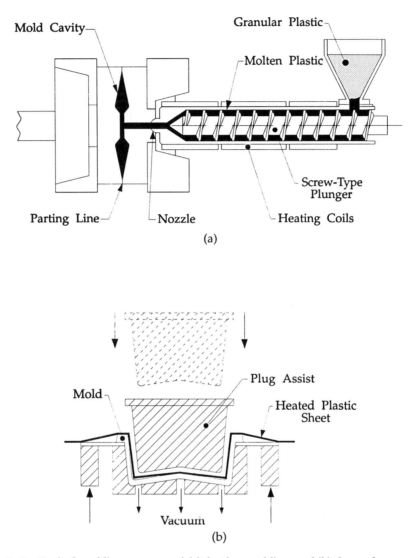

Figure 5–5. **Typical molding processes: (a) injection molding, and (b) thermoform molding.**

Process Mechanics and Process Conditions
Typical stages in plastic molding processes are as follows:

1. Mold or die system design, production, and materials
2. Plastification
3. Injection

4. Hardening or solidification

5. Ejection

Each of these steps will be described briefly.

Mold or Die System In general, mold or die systems must be able to do the following:

* Withstand the injection pressure (machines with a sufficient closing force)
* Withstand the required temperatures and temperature changes
* Conduct heat from solidifying material
* Withstand harsh chemical corrosion (acids)

Mold or die design mainly depends on the following:

* Type of material
* Temperature
* Pressure needed

Two major material groups, thermoplastic materials and thermoset materials, have different processing requirements. Recently, it has become possible to control the curing of thermoset materials so that more accurate processes can be used. Table 5–3 shows a rough classification of the two groups.

Mold materials for Group A processes are typically tool steels (highest requirements are for injection molding and extrusion) and alloyed steels. In Group B, typical mold materials include mild steel, aluminum, epoxy, wood, and plaster.

Plastification To prepare the plastic material for the forming process, it is necessary to distinguish between thermoplastic and thermoset materials. Figure 5–6 shows how thermoplastic materials behave when temperature is increased. The behavior is the reverse in thermoset plastics.

The main parameters that control the plastification process are the plastic

Table 5–3. Classification of molds/dies according to injection pressure/forming forces

Group A High injection pressure injection molding	Group B Low injection pressure casting/molding
Extrusion	Open-mold forming (reinforced materials)
Compression molding	
Transfer molding	Closed-mold forming (reinforced materials)
Blow molding	

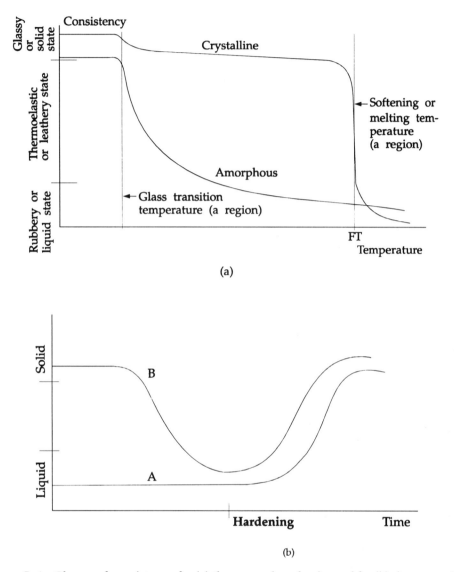

Figure 5–6. Change of consistency for (a) thermosetting plastics and for (b) thermosetting plastics during forming time for a liquid raw material and a granular (solid) raw material.

material, temperature, and pressure (for extrusion and injection molding). Often granular material is fed through a hopper into a heating zone and pressurized by a screw or plunger that creates a rubbery or liquid state. The material is injected into the die (Figure 5–5a) in this state.

For thermoset plastics, liquid or granular raw materials can be used (Figure 5–6). Forming must be terminated before final curing takes place. For liquid raw materials, casting is used; for granular raw materials, compression and transfer molding are used. Heating softens the material, and curing time is shortened. The curing process is irreversible.

Injection Depending on the specific material and the process, the injection (pouring, placing, or spraying) method is selected. For injection molding and extrusion, the injection pressure is rather high (between 7 and 28 ksi). This high injection pressure enables fine details and better dimensional tolerances to be achieved. Gating plays the same role as in metal casting, and gating design rules are highly developed. For most of the other processes, injection of the granular or liquid material is carried out without pressure (placing or spraying). Careful design of the injection method is a prerequisite for a good, sound product.

Solidification or Hardening Solidification or hardening (curing) takes place in the mold or die. For injection molding, which can also be used for thermoset materials, solidification takes place under pressure, and the die system is water-cooled to assist in shortening the process cycle time. In extrusion molding, cooling is introduced in the die and after the die to help create rapid solidification of the extruded material.

In general, temperature is the controlling parameter for thermoplastic materials, whereas time is the controlling parameter for thermoset materials. If solidification occurs under pressure, the highest accuracies are achieved. In plastic molding, just as in metal casting, shrinkage takes place during solidification.

Ejection When the material is solidified or cured, it can be removed from the mold. Often, the component is removed before it is completely solid to ensure a high production rate. In some processes, the ejection or removal of components is automatic, and in others, it is manual.

Process Conditions
The parameters that influence molding include the following:

  * Type of plastic
  * Type and amount of reinforcing material
  * Temperature of plastic
  * Injection pressure and time
  * Curing time (solidification)
  * Cooling (temperature of mold/die)
  * Pressure during curing
  * Design of mold

Workpiece Materials

Two major groups of plastic materials are as follows:

* Thermoplastic materials (reversible—can be softened or hardened by increasing or decreasing temperature)
* Thermoset plastics (irreversible—once hardened or cured, they cannot be softened again)

What to Expect from Components Produced by Molding

Thermoplastic or molded components have the following characteristics:

* Complex geometries can be produced
* High strength-to-weight ratios
* Usually electrically nonconductive
* Corrosion resistant
* Low weight
* Low unit costs
* Many functions can be built in

5.4 Lamination Processes with Composite Workpiece Materials

Lamination processes, which are common in the manufacture of composites, include the following:

* Filament winding
* Sheet laminating
* Bulk laminating
* Pultrusion

The primary focus of this section will be filament winding.

Filament Winding

Filament winding has the following characteristics:

* Uses a continuous length of fiber strand, roving or tape
* Produces a shell workpiece with a high strength-to-weight ratio
* The workpiece requires thermal energy for curing
* Winding patterns may be longitudinal, circumferential, or helical

Process Mechanics and Process Conditions

In filament winding, a continuous tape or roving of fibers, impregnated with resin, is mechanically wound over a mandrel to form the part. Successive layers are wound at the same or different angles until the required thickness is achieved (Figure 5–7). After the part is wound, it is left on the mandrel and placed in an autoclave to be cured with high compaction, or it is oven-cured when high compaction results are not required.

The relative motions between the mandrel and the head of the machine play an essential role in the success of the product. These motions determine the angle at which the filament is wound over the mandrel. The angle is predetermined through careful planning and calculating. The filaments can be arranged in various directions to achieve different strength properties. Two types of winding patterns are generally used:

* Hoop, circumferential, or radial
* Helical, axial, or longitudinal

In the hoop winding pattern, fibers are wound perpendicular to the axis of the mandrel (Figure 5–7) and the fiber strength is oriented in the radial direction. In the helical winding pattern, the fibers are oriented axially and have higher strength in the longitudinal direction.

The constant winding angle θ is defined by the following equation when both end openings (the holes where the mandrel passes through) are equal:

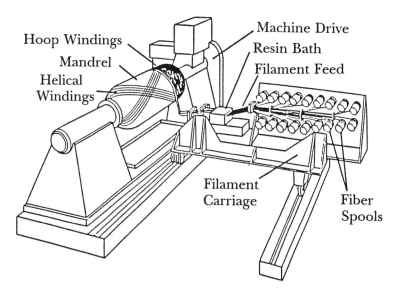

Figure 5–7. Common setup for a filament winding machine.

$$\sin \theta = \frac{polar\ diameter}{maximum\ diameter}$$

The wound part shape can either take on the shape of the mandrel or be pressure-formed during curing. The result of filament winding is a part made of a composite material. Various resins and filament materials can be used as the matrix and the reinforcement, depending on the desired properties of the part.

Resin Materials Resins are used as the matrix in filament-wound composites. Good resins have the following characteristics:

* Low in volatile content (preferably 100% solids)
* Viscosity is between 350 and 1500 mPa-sec
* Nonreactive diluents (used to lower viscosity) are avoided
* Pot life of several days
* Gelation of the resin does not occur before completion of winding
* The resin must flow during curing

Common resin materials include epoxy, polyester, phenolic, some imides, silicone, and thermoplastics. Polyesters and epoxies are most commonly used.

Reinforcement Materials The most common continuous reinforcements used for composites in filament winding are the following:

* E-glass (for lower costs)
* S-glass (for strength)
* Carbon (for strength and high stiffness [modulus])
* Aramids (for toughness and light weight)

Fibers process better if they are procured as untwisted fibers from the manufacturer and are well collimated.

Mandrel Materials If the part to be produced is open-ended, a simple mandrel cylinder of cored or solid steel or aluminum can be used. If the part to be produced is close-ended, special mandrels that can be removed while keeping the ends intact are required. If there is a large enough opening in the end, collapsible (segmented) metal mandrels are used. If the openings are small, several options are available. These options include using an inflatable mandrel, sand with a water-soluble binder [generally, polyvinyl alcohol (PVA)], soluble salts, eutectic salts, or low-melting alloys, which can be melted and removed.

Characteristics of Parts Produced by Filament Winding

Filament winding is a flexible method of manufacturing composite parts because of the ability to wind in one direction or at one angle and then change to another angle. It is often easy to join composite and metal or other materials. Parts with diameters ranging from 1 in. to 20 ft are commonly made by filament winding, depending on the machine size and capabilities. Composite materials can generally be used for high strength-to-weight ratios and some high temperature applications. Void presence of pressure-cured parts can be as low as 1%.

Deformation Processes

6.1 Introduction

Mass-conserving processes based on deformation have the following characteristics:

* The workpiece material is solid
* Processes are mass-conserving (dM = 0)
* Basic processing energy is mechanical
* Mechanical energy causes plastic deformation of the workpiece material
* The desired shape or geometry is created through relative motions between the workpiece material and the die, which can contain various geometrical patterns
* The workpiece material is malleable (deforms without fracture)

In deformation processes, the workpiece material is plastically deformed and takes on the geometric shape of the die. Deformation of the workpiece occurs at a temperature below its melting point, and the workpiece is never in a molten state. The basic process energy that causes the workpiece deformation is mechanical, and the deformation is caused by mechanical forces developed between the workpiece and the die. The desired shape is created through relative motions between the die and the workpiece material. No material is removed. The ability of the workpiece material to deform plastically without fracture is an important process parameter. The flow or deformation of the material depends on the material properties, the state of stress in the material, the temperature, and the strain rate of the workpiece material (velocity of deformation) during the process.

The following are important considerations of mass-conserving processes based on solid workpiece materials:

* State of stress in workpiece and process mechanics
* Process conditions
* Workpiece material
* Tool or die motions and materials
* Other workpiece characteristics

6.2 State of Stress

The stress under which deformation occurs plays an important role in the process mechanics, the tool design, and the equipment requirements. To study the processes in detail requires a subdivision in the major groups of stress states. The four types of stress states involved in mass-conserving processes are the following:

* Compression
* Tension (drawing or stretching)
* Bending
* Shearing

These general states identify only the dominant stress situations. Other stresses may also be present in the material along with the dominant stress.

Compression as the Dominant Stress

When the dominant state of stress in the process is compression, the tool or die geometry and the pattern of motions between the tool and the workpiece impose this stress. Processes in which compression is the dominant stress include drop forging, thread rolling, and upset forging (Figure 6-1).

A dominant state of compression can be illustrated by showing the deformation of a square block of material (Figure 6-2). The grid shows how the individual elements deform under the given conditions. For different geometries, the flow pattern varies. Even if compression is the dominant stress, other stress types are also present. In Figure 6-2, compression creates compressive strains in the vertical direction and tensile strains in the horizontal direction of the material. Fracture may occur if the ductility of the workpiece material is exhausted.

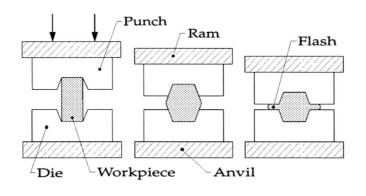

(a)

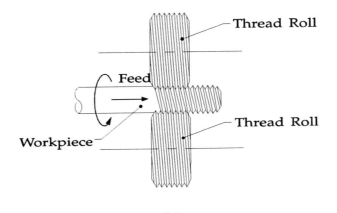

(b)

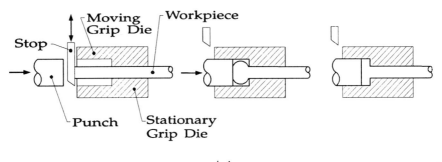

(c)

Figure 6–1 Processes with compression as the dominant stress: (a) drop forging, (b) thread rolling, and (c) upset forging.

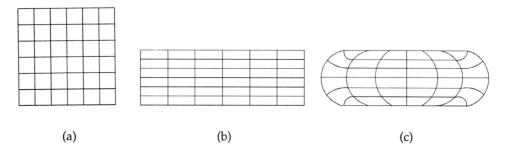

(a)   (b)   (c)

Figure 6–2 Deformation of square block by compression: (a) before deformation; (b) after deformation without friction between the workpiece and the die; and (c) after deformation with constraining forces (friction) resulting in barreling. The volume of the block is conserved after deformation.

A *stress–strain diagram* (Figure 6–3) for a given material is derived from a test that measures the stress and strain relationship of a material during elastic deformation, plastic deformation, and final fracture. These tests are carried out for tension, compression, torsion, and shearing. For example, the forces necessary to compress a cylindrical metal component to a certain height or strain level can be derived from the stress–strain diagram for the material. The stress–strain curve for compression is similar to the stress–strain curve for tension. Since the metal can sustain more plastic deformation in compression than in tension, the compression curve will show a larger region of plastic deformation. Figure 6–3 shows a compression stress–strain diagram for a hollow aluminum alloy cylinder under

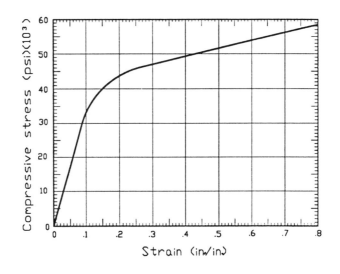

Figure 6–3. Compression stress–strain diagram for an aluminum alloy cylinder under well-lubricated conditions.

well-lubricated conditions. The stress shown is the true stress. The strain is the so-called natural, or logarithmic strain.

The plastic deformation portion of the stress–strain curve is often expressed by the equation $\sigma = K\varepsilon^n$, where σ is the true stress, K is a constant known as the *strength coefficient*, ε is the strain, and n is a strain-hardening exponent or coefficient. A large value of n indicates that small deformations in the material cause large increases in material strength due to cold working.

Tension as the Dominant Stress

Different materials have different tensile stress–strain curves (Figure 6–4). The shape and level of the curves depend on the particular material (alloy), the temperature, and the strain rate. Increasing temperatures lower the curve, and very high strain rates raise the curve.

In actual situations, the stress–strain curve for a particular workpiece material must be obtained to determine the correct process. Section 6.3 lists formulas to calculate stresses and strains.

In drawing or stretching processes, tension is the dominant state of stress. Examples of these processes include stretch forming, tube drawing, and conventional deep drawing (Figure 6–5).

Stretch forming occurs when a workpiece material is stretched over a die. In tube expansion, pressure is created inside a tube by liquid, elastic, granular, or rigid media, causing outward radial tensile stresses in the tube. In tube drawing, a tube is stretched, or drawn, over a floating mandrel. The tube thickness and diameter generally decrease, resulting in tensile stresses in the horizontal direction and compressive stresses in the vertical direction. In deep drawing, shrink

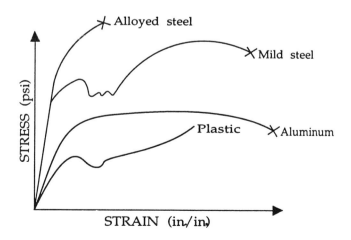

Figure 6–4. Tensile stress–strain diagrams for various materials.

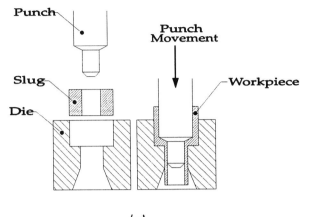

(a)

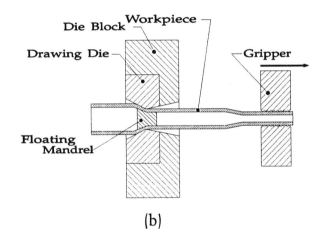

(b)

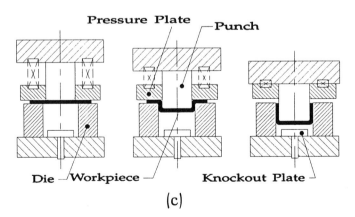

(c)

Figure 6–5. Examples of processes with tension as the dominant stress: (a) impact extrusion, (b) tube drawing, and (c) deep drawing.

forming and stretch forming can occur. When shrink forming occurs, a blank is pulled into smaller diameters, and the length generally increases. When this happens, compressive stresses are created radially from the circumference of the blank, which leads to buckling if a blank holder is not applied. When the blank is stretch formed, the diameter of the blank increases, and its thickness decreases. The dominant stress state in this case is tension. Drawn products are normally characterized by both shrink and stretch forming. The limitation in draw or stretch forming is local instability (cracking due to tension) of the metal. In compression, force is the limitation, as well as some instability created by the secondary stresses (usually tension).

Figure 6–6 shows, in principle, how a tension test is performed and the various stages of deformation represented on the stress–strain curve. Often, the nominal stress–strain curve is derived first. It is defined by the following equations:

$$\sigma_{nom} = \frac{P}{A_o} \tag{6-1}$$

σ_{nom} = nominal stress (psi)
  P = applied load (lb)
  A_o = initial cross-sectional area (in.²)

$$e = \frac{l - l_o}{l_o} \tag{6-2}$$

e = strain, or elongation (in./in.)
l = current length of test specimen (in.)
l_o = initial length of test specimen (in.)

Nominal stress–strain curves are often converted into true stress–strain curves. True stress–strain curves are useful in explaining the fundamental behavior of material. True stress (6–3) is determined by dividing the load by the actual cross-sectional area of the test specimen during the test.

$$\sigma_{true} = \frac{P}{A_{act}} \tag{6-3}$$

True strain is computed by using the actual length of the specimen during the test.

$$\varepsilon = \ln\frac{l}{l_o} = ln(e - 1) \tag{6-4}$$

The difference between nominal and true stress and strain becomes important as stresses and strains become large. Only true stress–strain curves can be used with plastic deformation. Instability occurs from uniaxial tension in the work-

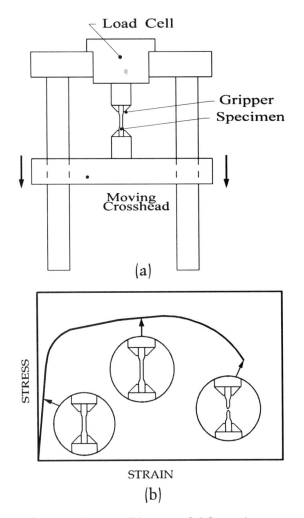

Figure 6–6. (a) Setup for a tension test, (b) stages of deformation represented on a stress–strain diagram.

piece when the strain is equal to the deformation hardening exponent ($\varepsilon_{\text{instable}}$ = n). If tension is applied in two directions, $\varepsilon_{\text{instable}}$ equals 2n.

Bending as the Dominant Stress

Bending processes are characterized by compression at the inside of the bend and tension at the outside of the bend. In Figure 6–7, σ_c represents the maximum compressive stress at the top surface of the beam, and σ_t represents the maximum tensile stress at the bottom of the beam. F_c is the compressive force in the top portion of the beam, and F_t is the tensile force in the bottom portion of the beam. Based on assumptions of elastic and plastic deformation in beams, and the posi-

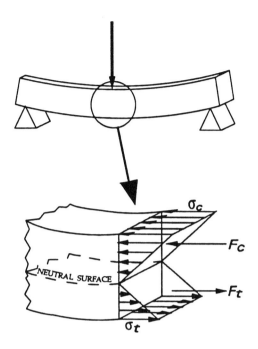

Figure 6–7. Stress distribution in bending.

tion of the neutral surface in the beam, the strains on the top and the bottom of
the beam can be calculated; the minimum acceptable bending radius can also be
calculated. Tube bending and roll bending are examples of bending processes
(Figure 6–8).

Bending can be done with or without backing dies, depending on the required
size of the bending radii. Bending is rather difficult to predict theoretically, so
empirical data are important.

Shearing as the Dominant Stress

Mass-conserving processes rarely involve shearing as the dominant stress. The
most important processes utilizing the shear mechanism are shear spinning and
torsion (Figure 6–9).

Shear properties of a material are usually derived from a torsion test, as
illustrated in Figure 6–10. The torsional stress–strain diagram has the same general shape as the tension or compression stress–strain diagram. The stress in this
case is the torsional shear stress (τ), which is represented as follows:

$$\tau = \frac{T}{2\pi r^2 t} \tag{6–5}$$

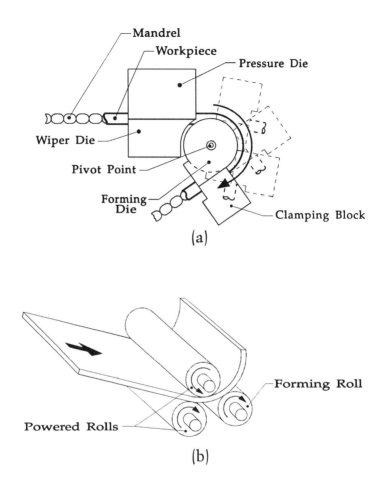

Figure 6–8. Examples of bending processes: (a) Tube bending and (b) roll bending.

T = torque (in.-lbs)
r = radius of the gage section
t = thickness of the gage section

The strain in this case is shear strain, which is represented as follows:

$$\gamma = \frac{r\phi}{L} \qquad\qquad (6\text{–}6)$$

r = radius of test specimen (in.)
ϕ = shear angle
L = length of test specimen (in.)

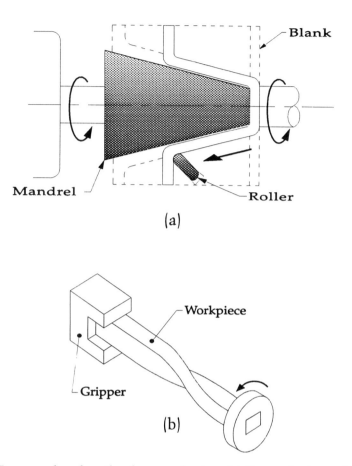

(a)

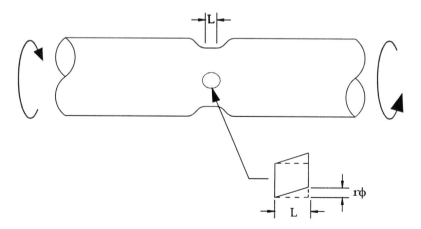

(b)

Figure 6–9. Processes based on the shear mechanism: (a) Shear spinning and (b) torsion.

Figure 6–10. Torsion test. The workpiece is a thinwalled tube and the torsion is assumed to take place entirely in the reduced diameter gate section.

By using the yield criteria, the shear situation can be compared with the tension and compression situations.

6.3 Process Conditions

Process conditions describe or determine the conditions under which a process is carried out. In general, to ensure a satisfactory product, the process conditions must be studied to determine equipment parameters (force, work, speed, etc.), tool geometry, preform of workpiece material, lubrication, and temperature. The following parameters are important in deformation processes:

* Tool geometry
* Workpiece material preform
* Workpiece material temperature
* Force, work, and velocity of deformation
* Friction and lubrication

Tool Geometry

Tool geometry must be carefully designed. Radii, draft, flash, and other dimensional parameters must be dimensioned according to general theoretical or practical rules. Tool geometry parameters depend on the specific workpiece material, the preform shape, and the required workpiece deformation.

The *Manufacturing Processes Reference Guide* gives guidelines for tool and die design and workpiece material preform. The preform is important to ensure proper filling of the die and to eliminate fracturing in the workpiece material. Based on theoretical calculations of strains, suitable preforms can be determined. It is also possible to simulate a process by using wax materials and wooden dies.

Workpiece Material Temperature

If the temperature is raised above the workpiece recrystallization temperature, the yield stress of the workpiece decreases considerably, and large strains in compression can be obtained. This method is heavily utilized in forging, rolling, and extrusion. The elevated temperature produces a rough surface due to scaling. A low yield stress enables lighter equipment to be used. Hot forming and cold forming processes are distinguished by temperature.

Force, Work, and Velocity of Deformation

To select equipment and to dimension tools or dies, it is necessary to calculate the required force, the work loads, and the velocity of deformation. Several methods with varying degrees of accuracy are available. Some are theoretical, and some are empirical. It is very important that engineers have a fundamental knowledge of the mechanics of metal forming so they can evaluate the data for different deformation processes.

The necessary work for a certain plastic deformation can be calculated from the following equation:

$$dW = \sigma' \times d\varepsilon' \times dV \qquad (6\text{--}7)$$

dW = incremental work of deformation (lb-in.)
σ' = equivalent stress (psi)
$d\varepsilon'$ = incremental equivalent strain (in./in.)
dV = volume element of deforming body (in.^{3})

If the stress–strain curve for the material can be represented by the model $\sigma' = k\varepsilon'^n$, and if the various elements in the deforming body get the same deformation, the following equation is obtained by integrating (6–7):

$$W = V \int_0^{\varepsilon'_f} k\varepsilon'^n d\varepsilon' \qquad (6\text{--}8)$$

or

$$W = V \frac{k}{n+1} \varepsilon'^{n+1}_f \qquad (6\text{--}9)$$

If the necessary force is to be calculated, the following equation can often be used:

$$P_{average} = \frac{W}{l_{average}} \qquad (6\text{--}10)$$

where $P_{average}$ is the average force, and $l_{average}$ is the average deformation length. For example, if a simple compression operation is carried out, the force can be calculated by the following equation:

$$P = \sigma' A \qquad (6\text{--}11)$$

σ' = equivalent stress
A = actual area

The effective or *equivalent stress* is based on the *von Mises' yield criterion* and is given by the following equation:

$$\sigma' = \left[\frac{(\sigma_1 - \sigma_2)^2 + (\sigma_2 - \sigma_3)^2 + (\sigma_3 - \sigma_1)^2}{2} \right]^{1/2} \qquad (6\text{--}12)$$

σ_1, σ_2, and σ_3 ($\sigma_1 > \sigma_2 > \sigma_3$) are the principal stresses.

The *equivalent strain* is given by the following equation:

$$\varepsilon' = \left[\frac{2}{3} \left(\varepsilon_1^2 + \varepsilon_2^2 + \varepsilon_3^2 \right) \right]^{1/2} \qquad (6\text{--}13)$$

ε_1, ε_2, and ε_3 are the principal strains.

For bending and shearing, more specific formulas have been developed.

Friction and Lubrication

Friction under high pressure plays an important role in deformation because it influences the necessary force and flow and the resulting surface quality of the workpiece. Therefore, lubricating to reduce friction is extremely important. Several lubricants have been developed to withstand rough workpiece conditions. For each process, the lubricant must be carefully selected.

6.4 Workpiece Materials

A close relationship exists between the workpiece material, the deformation process, and the workpiece geometry. This means that a material must possess certain properties in order to be used in a specific process or process family. It is absolutely necessary to know the properties of a specific workpiece material in relation to a specific family of processes being considered for a manufacturing operation. For example, a brittle material cannot be forged.

The properties of a material depend on several parameters, including the state of stress and strain in the material, the temperature at which the deformation process is carried out, and the velocity of deformation. Hydrostatic pressures generally increase the formability of a material. Often, materials have been developed for certain process families; selecting these materials is recommended.

Compression Processes

Shaping in compression requires that the material has good formability (high ductility under the given process conditions). Formability can be generally increased in workpiece materials by heating above the recrystallization temperature or by introducing hydrostatic pressure.

Limitations in compression processes include fracture of the workpiece material and the required amount of force to produce the desired deformation. As mentioned, at high temperatures, large strains can be obtained in compression.

The process speed also has some influence on formability. The higher the speed, the lower the formability, especially at high temperatures.

Drawing Processes

The workpiece must be able to withstand the required deformation by drawing and stretching in tension. This is a harder requirement to meet than in compression. The deformation is normally limited by instability, and, depending on the state of stress and strain in the workpiece, the limit occurs when $\varepsilon_{instable}$ is between n and 2n, where n is the deformation hardening exponent. In tension, the formability is decreased by higher temperatures and speeds.

Bending Processes

For bending processes, good ductility is required, and the processes are limited by tensile stresses in the workpiece. Tensile stresses cause workpiece instability or fracture. Different materials have different properties in bending. Empirical data are listed in reference books.

Shearing Processes

In general, high ductility is desired, and workpiece instability and fracture limit the use of shearing processes. Increased temperatures generally increase ductility and resulting formability.

6.5 Tool/Die Motions and Materials

As discussed previously, the shape of the component is created by the interaction of a die, which contains the desired workpiece geometry, and the workpiece material. The interaction is described by relative motions or kinematic patterns. Four situations or types of interaction between a die and a workpiece can be identified:

* **Total forming,** where the die contains the whole desired geometry and only a closing motion is required (e.g., forging; see Figure 6–1).
* **One-dimensional forming,** where the die contains only a producer so that one relative motion is required between the die and the material (e.g., thread rolling; see Figure 6–1).
* **Two-dimensional forming,** where the die or tool affects only a small portion of the desired workpiece surface, and two relative motions are required (e.g., shear spinning; see Figure 6–9).

* **Free forming**, where the die or tool does not contain any part of the desired workpiece surface, and the imposed stress situations create the shape of the workpiece (e.g., free upsetting; see Figure 6–1).

The less surface contour in the die system, the more complicated the pattern of motion becomes. With decreasing intricate surface contents in die systems, smaller forces are necessary to deform the workpiece (see Figures 6–11, 6–12, and 6–13 for examples of tool die motions in deformation processes).

The pattern of motion of a deformation process depends on equipment capability. Often a tool or die is designed based on the pattern of motion available in the particular equipment. For example, the selection of the tool or die material depends on the load to which it is subjected. The *Reference Guide* lists typical

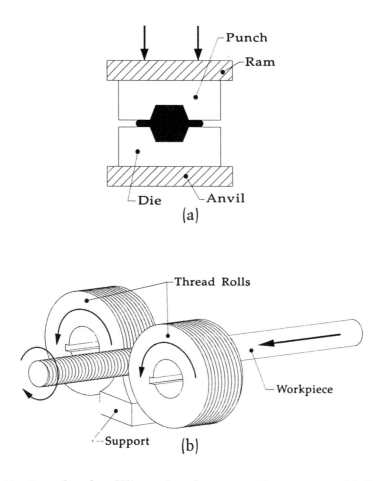

Figure 6–11. Examples of tool/die motions for compression processes: (a) Drop forging and (b) thread rolling.

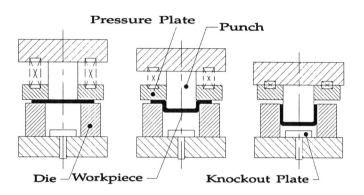

Figure 6–12. Example of tool/die motions for drawing/stretching processes: Deep drawing.

tool or die loads and the corresponding possible tool materials for the four subgroups of deformation processes.

6.6 Characteristics of Components Produced by Deformation Processes

It is possible to list what to expect from deformation processes, but first it is necessary to distinguish between hot and cold forming, because they determine the resulting surface finish and tolerances of the workpiece.

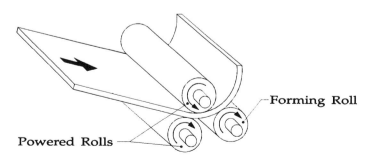

Figure 6–13. Examples of tool/die motions for bending processes: Plate roll bending.

Hot forming takes place at temperatures above the workpiece material recrystallization temperature and includes rolling, forging, and extrusion. Due to the lowered yield strength in hot forming, the necessary forces and power are much smaller than in cold forming. Hot forming processes produce the following:

* Relatively rough surfaces due to scaling
* Relatively wide tolerances (2–5%) due to the rough surfaces
* A wide variety of shapes (large deformations are obtainable in compression)
* A refined material structure resulting in good mechanical properties (excellent ductility, impact strength, and generally good fatigue resistance).

Cold forming takes place at temperatures below workpiece recrystallization temperatures; it includes rolling, extrusion, forging, deep drawing, and bending. Compared to hot forming, cold forming processes produce the following:

* Better surfaces and tolerances
* Better mechanical properties (strength and fatigue resistance)
* Anisotropy (i.e., directional properties of material; this is only an advantage when it is possible to utilize this effect, as in deep drawing)
* Less ductility in the work material (due to strain hardening)

In general, the original workpiece material must have a clean, scale-free surface. The forces required to carry out cold forming processes are high due to the relatively high yield strengths. This often results in high die loads.

Mechanical, Thermal, and Chemical Joining Processes

7.1 Introduction

Joining is listed in the taxonomy chart as a branch of shaping processes, along with mass-conserving and mass-reducing (Chapter 1). Producing final shapes by joining subcomponents may also be referred to as mass-increasing. Both permanent and nonpermanent joints may be used to join subcomponents. In general, permanent joining processes have the following characteristics:

* They create permanent joints
* Mechanical, thermal, or chemical energy is utilized to create joints
* Filler materials are sometimes necessary to create sound joints

Joining is divided into three subfamilies according to the type of energy used: mechanical, thermal, or chemical. In *mechanical joining*, high pressures or forces create the joint. In some processes, high pressure is used to plastically deform the surfaces to be joined. This deformation helps remove oxides and expose virgin metal to create a strong joint. The processing temperature of mechanical joining processes is less than the workpieces' melting points. *Thermal joining* includes processes that involve localized melting of the material, allowing surfaces to fuse together. Sometimes a filler material is used to help create the joint. Thermal joining processes are characterized by high temperatures. In *chemical joining*, chemical adhesives create the joint.

7.2 General Process Mechanics

A permanent joint is defined as a localized union between the elements to be joined, based on either cohesion, adhesion, or both. In cohesion (the attraction between molecules that holds a mass together), the appropriate combination of temperature and pressure causes the formation of a common metallic structure in the joint. This type of joining requires that the elements to be joined have the same basic structure and composition. In adhesion (the attraction of unlike molecules between surfaces), the elements are bonded together through physical, electrical, or chemical surface forces; there are no common structures formed. Often, adhesion is actually created between the components and a filler material, not between the components themselves.

To obtain a sound joint based on coalescence, the following must occur:

1. The surfaces must be free of oxide layers, absorbed gas, and other contaminants

2. The surfaces to be joined must be placed together so that the bonding forces can be activated

Each joining process that relies on coalescence meets these requirements differently.

7.3 Mechanical Joining

Process Mechanics

Mechanical joining is based on coalescence, which is caused by the combination of pressure and temperature. Coalescence is the union formed when the surfaces to be joined actually merge together. These processes are characterized by high pressures; temperatures are always below the melting point of the materials being joined.

There are two types of mechanical joining. The first type relies on mechanical means to cause the pressure that creates the plastic flow. The second type of mechanical joining involves heating the material to lower its yield stress and facilitate the joining process. This second type of mechanical joining may use mechanical or chemical energy to increase the temperature. Both types of mechanical joining bring the surfaces close together so that joining may occur; they also provide a way to remove oxides.

Figure 7–1 illustrates three typical mechanical joining methods. Figure 7–1a shows cold welding in which pressure and motion between the two components create the welding conditions. Figure 7–1b shows friction welding in which heat is created mechanically. Finally, Figure 7–1c shows explosive welding in which the mechanical energy is created by explosives.

Process Conditions and Work Materials

Each mechanical joining process has different conditions, as listed below:

Cold Welding
  * Pressure
  * Sliding between the shear faces (occurs when there is surface deformation)

Friction Welding
  * Pressure (force)
  * Rotation (torque)
  * Temperature

Usually, work materials that are mechanically joined must be electrically conductive. In general, the following major material groups are exposed to these processes:

  * Steels (mild and alloyed)
  * Aluminum and aluminum alloys
  * Copper and copper alloys

Characteristics of Mechanical Joining

Components produced by mechanical joining methods often require some additional machining or trimming operations. The tolerance level obtained depends on the fixture design. The better the fixture design, the better the accuracy obtained. The strength properties are not as good as the base material. Many of the mechanical joining processes are highly automated.

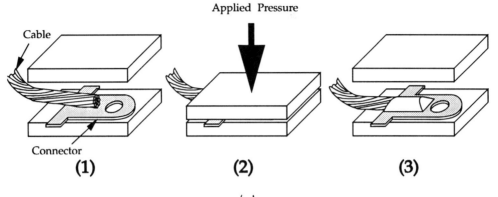

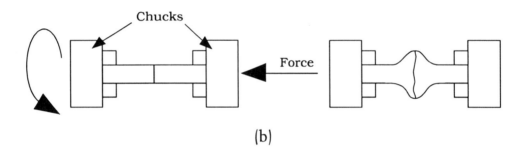

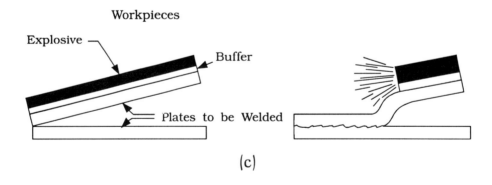

Figure 7–1. Mechanical joining processes: (a) Cold welding, (b) friction welding, and (c) explosive welding.

7.4 Thermal Joining

Process Mechanics of Thermal Welding

As mentioned, thermal welding is a category of permanent joining; it can be defined by the following characteristics:

* The melted material flows together and the joint is formed by cohesion
* The thermal energy required to melt the material is created from chemical or electrical energy
* When filler material is used, it must have the same basic composition as that of the materials being joined

Heating metal allows for accelerated oxidation. Therefore, in thermal welding, the joint must be protected or shielded from oxidation and other contaminants (Figure 7–2).

The basic elements of a thermal welding setup are listed below:

* Heating source
* Filler
* Shielding/protection

Each of these three elements is provided in different ways. For example, the heating source could be an electric arc, a flame, a laser, or an electron beam. In resistance welding, the heat is created by electrical resistance in the gap between the components to be joined. Filler materials can either be the electrode of an electric arc welding process or a separate filler wire or rod. Even shielding

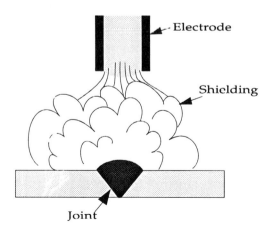

Figure 7–2. Shielding of the molten pool of metal in thermal welding.

Table 7–1. Classification of heating source, electrode, filler, shielding/protection, and process for thermal welding

Energy source	Type of electrode	Source of filler material	Shielding	Welding process
Electric arc	Consumable	Electode	None	Metal electrode arc welding
Electric arc	Consumable	Electrode	Electrode coating	Shielded metal-arc welding
Electric arc	Consumable	Electrode	Granular powder	Submerged-arc welding
Electric arc	Consumable	Electrode	Stable/inert gas (CO_2, He, Ar)	Gas metal-arc (MIG welding)
Electric arc	Nonconsumable	Separate filler	Stable/inert gas (CO_2, He, Ar)	Gas tungsten-arc (TIG welding)
Flame	Not applicable	Separate filler rod	The flame	Gas welding

methods vary. Gas, powders, electrode coatings, and gas flames are all used as shielding methods, according to the type of process. Each type of thermal welding has its own name. These variations of thermal welding are outlined in Table 7–1.

In Figure 7–3, several process schematics illustrate the differences among thermal welding processes.

Electrodes, Fillers, and Coatings

If an electrode is consumable, its composition must be similar to that of the material being joined, since it also acts as a filler. It should be noted that welding electrodes made of tungsten are nonconsumable and require that an additional rod or wire of some other material be used as the filler. Often, coated electrodes are used. Typical coating materials include calcium oxide (CaO), calcium fluoride (CaF_2), and silicate (SiO_2). Powders of the metals to be welded may be included in the coating to give a higher productivity rate (amount of molten metal/unit of time). In processes where electrodes are not used, another type of filler is used, but for both electron and laser beams, filler materials are usually not necessary.

For best results, it is necessary to carefully select a shielding method. Different processes require different protection methods. In gas welding, the flame itself is the protection. In laser welding, a shielding gas is used, and electron beam welding is carried out in a vacuum.

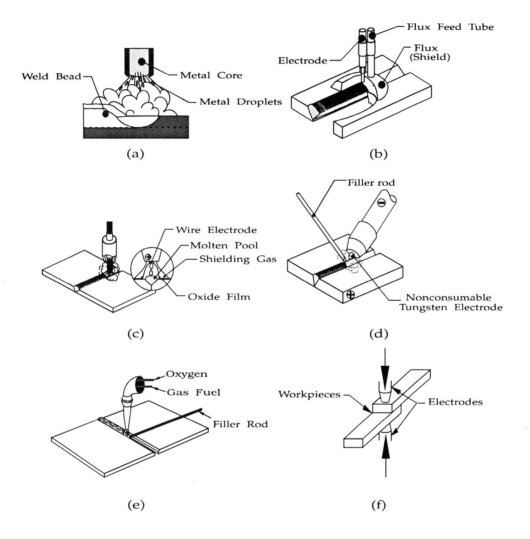

Figure 7–3. Thermal welding processes: (a) Shielded metal arc welding, (b) submerged arc welding, (c) MIG welding, (d) TIG welding, (e) gas welding, (f) spot welding.

Process Conditions and Work Materials

The main parameters for electric arc welding processes are the following:

* Type of material to be joined
* Shape of joint
* Potential difference between electrode and workpiece
* Distance between workpiece and electrode
* Type of filler material to be used
* Method of protection/shielding

Each of these parameters must be maintained within certain limits to achieve optimal results. For example, the potential differences between the electrode and workpiece should be between 18 and 35 volts during welding and between 50 and 80 volts when starting the arc. Selecting the appropriate shielding methods is also important. Gas is used as shielding for MIG and TIG welding. In shielded metal arc welding, the slag produced provides protection from oxidation.

Joint design also requires some consideration (Figure 7–4). To achieve acceptable joint quality, advance preparation is necessary. This is especially true for laser and electron beam welding. When lasers and electron beams are applied, butt joints or special joints are used due to the small dimensions of the beams. These processes require workpiece precision and proper fixturing to ensure a correct weld.

The type of material used also affects the outcome of welding processes. Most welded materials are mild steels, alloyed steels, aluminum alloys, nickel alloys, and copper alloys. It is important to ensure that a certain material can be joined by the thermal welding process chosen. Some processes work better on specific materials than others.

Each thermal welding process has specific characteristics. For example, electric welding processes are easily automated; automation of these processes is becoming much more common. Gas welding also has special considerations. In gas welding, which is used mostly for repair work, it is necessary to adjust the fuel correctly (oxygen and acetylene mixture) in order to create the proper flame. Also, when moving the flame, care must be taken to protect the molten metal adequately from oxidation. In addition, gas welding requires the appropriate selection of a filler metal. Resistance welding also has unique characteristics. In this process, thermal energy is created electrically by resistance heating the components. This type of welding is possible because the resistance is greatest at the gap, so the greatest amount of heat occurs there. Processes that use resistance heating include spot welding and flash butt welding. The parameters that influence resistance welding processes are also different from other thermal processes.

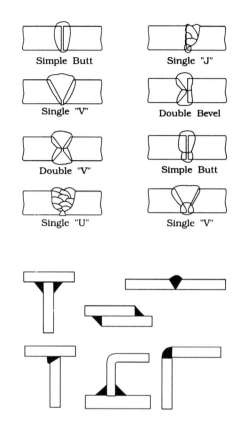

Figure 7–4. Examples of welded joints.

Voltage between the electrodes, pressure between the electrodes, and timing must all be selected carefully to ensure good results.

Characteristics of Thermal Welded Components

With the right conditions and adequate preparation, a weld with very good properties can be produced. Often, the weld has properties very similar to those of the materials being joined. One drawback of thermal welding is that close tolerances are difficult to achieve, so these processes usually require additional machining. Also, if special accuracy or stability is required, stress relieving or annealing before machining may be necessary.

If laser or electron beams are used, welding may be the final process. All other machining can be done prior to the joining process. With these types of welding processes, only component accuracy and fixturing limit a part's dimensional accuracy (±0.0005 in.).

Resistance welding has its own advantages. The tolerances achieved in resistance welding processes depend on the quality of the fixture. Resistance welding processes are also easily automated, allowing for high production.

7.5 Chemical Joining

Chemical joining processes, more specifically known as adhesive bonding processes, have the following characteristics:

* They create a permanent joint
* Heat is sometimes used in the process
* Filler materials (chemical adhesives) are used to create the joint

Process Mechanics

Adhesive bonding is possible due to adhesive forces between the work materials and the adhesives. Adhesion is the force that binds unlike molecules between adjoining surfaces. Usually, the bonding is between the surface of the workpiece and the adhesive, not between the joining surfaces themselves. This principle allows for the joining of dissimilar materials. Adhesive bonding is characterized by the following:

* The joining is based on adhesion forces between a filler material and the workpiece materials
* The melting point of the filler material is generally lower than the melting point of the components to be joined
* The filler material can be one of many nonmetallic, chemical adhesives
* The basic steps are either mechanical (such as the application of filler into the joint) or thermal (such as melting, solidifying, and curing)

Bonding between parts is based on adhesion and on the diffusion of the filler material into the surfaces of the parts. In a liquid state, adhesives will flow by capillary forces into the joint geometry. To obtain a good joint, surfaces must be clean, joints must be shaped correctly, and the joining temperatures must be correct. Figure 7–5 shows typical joints for adhesive bonding.

Process Conditions

The main parameters in adhesive bonding are the following:

* Workpiece material
* Selection of adhesive

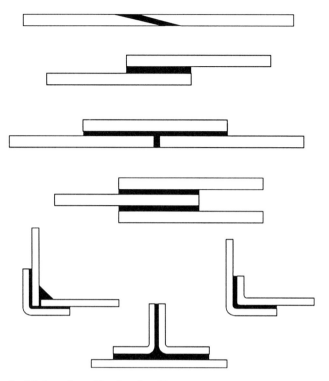

Figure 7–5. Typical joints for adhesive bonding.

* Joint geometry
* Joining procedure
* Temperature
* Curing time

In adhesive bonding, the joining procedure is especially important. This includes proper fixturing, correct application of force during curing, and adequate curing time. It is critical to follow carefully all recommendations from the adhesive manufacturer.

Workpiece Materials and Adhesives

In adhesive bonding, the filler material (the adhesive) has several components: a base material, a solvent, a filler material, and a hardener. The base material, which gives the adhesive the desired adhesion properties, is usually a thermoplastic or a type of thermosetting resin. The solvent gives a suitable viscosity to the adhesive. The filler material may be added to increase strength or reduce thermal expansion. The hardener is added to activate the adhesive.

Many adhesives are available for a variety of applications, so the specific material to be joined must be selected carefully. Also, the manufacturer's recommendations for the adhesive must be followed precisely. Adhesive bonding can be used for most work materials except cast iron.

Characteristics of Chemical Joining

Adhesive bonding has broad industrial applications and is expanding rapidly because many excellent new adhesives have been developed. Depending on the filler materials and the actual joints, a shear strength between 4000 and 40,000 psi can be obtained. The dimensional accuracy of a bonded part depends mainly on fixturing.

7.6 Bonding

Process Mechanics

Bonding is characterized by the following:

* The joining is primarily based on adhesion forces between a filler material and the workpiece materials; in some processes, such as brazing, cohesion also appears
* The melting point of the filler material is lower than the melting point of the components to be joined
* The filler material is usually metallic, as it is in brazing and soldering processes
* The basic steps involved are either mechanical (such as the application of filler into the joint) or thermal (such as melting or liquifying)

The bonding between the parts is based on adhesion and on the diffusion of the filler material into the surface of the parts. Sometimes diffusion is achieved when the filler material, in a solid state, is placed in the joint and heated. This allows the filler to melt and to flow into the surfaces to be joined. To obtain a good joint, surfaces must be clean, joints must be shaped correctly, and the joining temperature must be correct. Figure 7–6 shows typical joints for brazing and soldering.

Filler Materials/Workpiece Materials

The filler material to be used depends on which process is used. A wide variety of filler materials is available. Brazing and soldering use metallic, nonferrous

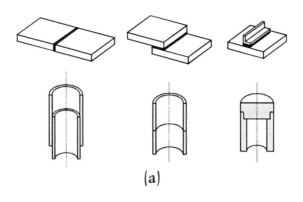

(a)

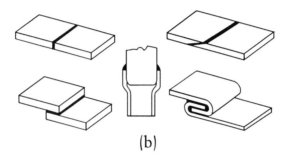

(b)

Figure 7–6. Typical joints for bonding processes: (a) Brazing and (b) soldering.

fillers. The melting point of fillers for brazing processes is usually above 850°F. These fillers include the following:

* Copper-based fillers alloyed with Ni, Co, and Cr
* Brass-based fillers alloyed with Zn, Mn, and Ni
* Silver-based fillers alloyed with Cu, Zn, and Cd
* Aluminum-based fillers alloyed with Si

Copper-based fillers are primarily used for joining steel and carbides; brass-based fillers are used for joining copper and copper alloys; silver-based fillers are used to join copper, copper alloys, and steel; and aluminum-based fillers are used for joining aluminum and aluminum alloys. In soldering, the melting point of fillers is usually below 850°F. The main fillers used are tin and lead alloys with small amounts of antimony. For soldering light metals, tin/zinc/cadmium alloys are used.

The process referred to as diffusion bonding does not use any filler material. Instead, the components are cleaned, clamped together, and placed in a vacuum furnace. The temperature creates material diffusion between the parts. The basic requirement is that the two parts have constituents that allow diffusion.

Process Conditions

The main parameters to achieve a good joint are the following:
* Work material
* Filler material selection
* Joint geometry
* Joining procedure
* Temperature

Characteristics of Bonded Workpieces

Bonding has broad industrial applications. Depending on the filler materials and the actual joints, shear strengths of between 4000 and 40,000 psi can be obtained. The dimensional accuracy of a bonded part depends mainly on fixturing. Most brazed and soldered components must be cleaned afterwards to remove the fluxes.

Annealing (Softening) Processes

8.1 Introduction

Occasionally, it is desirable to change the material properties of a workpiece. Thermal treatments can be used to alter a material's properties. This chapter will discuss softening, or annealing, processes, which are used to change the properties of different metals. During almost any metal manufacturing process, the structure and properties of the workpiece material are changed. Sometimes these changes are undesirable. For example, grinding operations on hardened steel create high stresses in the workpiece and may cause microcracks on the surface. Likewise, thermal hardening treatments may cause a workpiece to become brittle. Sometimes mechanical forming processes cause the workpiece to become brittle due to cold working. Many times these changes are unacceptable because they cause poor part performance.

Thermal softening treatments are utilized for a variety of reasons:

* They restore a workpiece's softness and ductility
* They reduce internal stresses in a workpiece without sacrificing significant hardness or strength
* They modify workpiece grain size
* They improve a workpiece's machinability

Different softening heat treatments have the same basic purposes, but slight variations exist. In this chapter, each process is characterized according to how the material's structure is affected. When a metal is heated, its structure undergoes three stages: recovery, recrystallization, and grain growth.

Recovery

When reheating previously cold-worked metal, the first noticeable effect is the reduction of locked-in internal stresses. This stage, known as recovery, reverses the effects of cold working. The increased temperature causes the atoms to move more freely and to approach their equilibrium position. Recovery is usually not detectable by a microscope, but evidence of recovery can usually be found by x-ray diffraction methods.

Recrystallization

At higher temperatures, the crystal structure starts to break down, and new and unstrained metal grains form. This new change, called recrystallization, is easy to detect because it has a marked effect on the properties of the metal (Figure 8–1). Recrystallization results in the formation of nuclei for new stress-free grains (Figure 8–2). Nuclei appear to form at regions where atomic structure is most highly disarranged. New stress-free grains grow from distorted grains, beginning at grain boundaries and along severely distorted slip lines and twin planes.

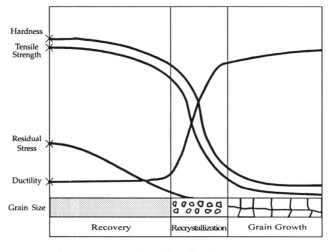

Annealing Temperature

Figure 8–1. Changes in properties as a result of annealing cold-worked metal.

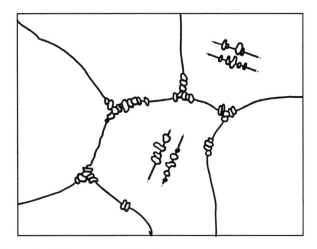

Figure 8-2. Schematic of the recrystallization process.

Grain Growth

When the temperature of cold-worked metal is kept above the recrystallization temperature, the new grains grow rapidly by absorbing other grains; this process is called grain growth. Because mechanical properties of metals are so closely related to grain size, control of grain size is important to the metallurgy of cold working and annealing. Although much research has been conducted, grain growth is not yet completely understood. However, it is generally recognized that the driving force for grain growth in a recrystallized metal lies in the surface energy of the grain boundaries. Small grains have higher surface energy than large grains. Surface energy is reduced as large grains grow larger and thus completely absorb smaller grains. In the process of grain growth, the total number of grains is reduced.

In this chapter, softening processes are classified into two categories: recovery and recrystallization. Recovery processes differ from recrystallization processes in that the temperatures used in recovery processes are not high enough to cause the material to recrystallize.

8.2 Definition of Terms

Phases

Many of the changes that a metal undergoes during thermal treatments are caused by changing the phase of the metal. Like most materials, metals have a liquid phase and a solid phase. Some metals, especially ferrous alloys, have several solid phases. Depending on the alloy of the metal and the applied heat treatments, metals can have one solid phase or a mixture of different phases. Many terms are used to describe these phases. One phase, known as martensite, is an unstable phase and can only be achieved in some materials by a rapid cooling technique known as quenching. Ferrite is a phase of low carbon steels. Austenite, another phase of steels, requires that there be less than 2.1% carbon content. Bainite, spheroidite, and pearlite are mixtures of one of the phases of steels and tiny carbide structures. In iron these carbide structures are also known as cementite. Sometimes pearlite is described as being lamellar, which means that it is composed of alternating layers of ferrite and carbides.

IT Curves

IT (isothermal transformation) curves or diagrams are used to determine temperatures and times for quenching and tempering processes. An IT curve is distinguishable by its shape, which is likened to a nose (Figure 8–3).

Process Names

Nearly all the thermal processes used to soften work-hardened materials or relieve internal stresses in a workpiece are referred to as annealing processes. Tempering processes, annealing processes that usually follow quench hardening, relieve internal stresses and increase toughness of quench-hardened workpieces. Tempering

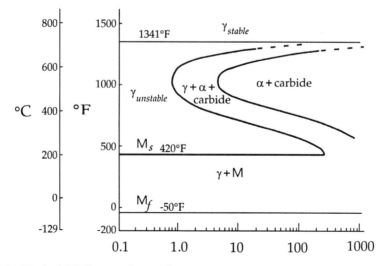

Figure 8–3. Typical IT diagram for steel.

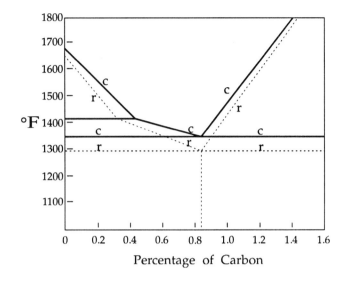

**Figure 8–4 Critical temperature diagram. Temperatures used for cooling ----------.
Temperatures used for heating _____.**

can also decrease workpiece hardness, reduce its strength, stabilize the structure, or change the volume of a workpiece.

Critical Temperatures

The temperature used in heat treatment processes depends on the critical temperature of the material being treated. Critical temperatures for a given material occur at the point where that material experiences a phase change. Figure 8–4 is a phase diagram (also known as a critical temperature diagram) for steel. Some diagrams show two critical temperatures at each phase change. This is due to the differences in temperature at which the phase change takes place, depending on whether the material is being heated or cooled. Figure 8–4 illustrates how critical temperatures vary, depending on whether the steel is being heated or cooled. The critical temperatures associated with cooling are marked with a dashed line and are labeled with lowercase r's. The critical temperatures for heating are labeled with lowercase c's.

8.3 Recovery Processes

Recovery processes are heat treatment processes in which the temperatures are not high enough to destroy the metal's granular lattice and cause recrystallization. Recovery processes include the following:

* Stress relieving
* Customary tempering
* Martempering
* Austempering
* Isothermal quenching

Stress Relieving

Stress relieving is an annealing process that is used to relieve internal stresses created in metal as a result of plastic deformation caused by forming, machining, and grinding operations. Stress relieving also helps remove residual stresses caused by uneven heating or cooling in weldments, castings, and forgings. These internal stresses may cause serious warping and even part failure. Internal stresses sometimes cause castings to warp even after they have been accurately machined and assembled into a complicated machine tool. If this warping occurs after the finished machine has gone into service, the machine may be inaccurate or inoperable. Stress relieving consists of heating a workpiece to the recommended temperature, holding it for a sufficient length of time so that it becomes uniformly heated throughout, and cooling it in air. Stress relieving treatments are often used after hardening operations because they can be carried out so that the material's strength and hardness are not substantially affected. Stress relieving operations conducted at lower temperatures (approximately 550°F) on cold-worked steel are sometimes called strain tempering, whereas those conducted at temperatures approaching the lower critical temperatures are sometimes called strain annealing. Stress relief treatments that have been recommended for ferrous metals are shown in Table 8–1. The effect of stress relieving at various temperatures is shown in Figure 8–5.

Tempering

The heat-treating operation known as tempering nearly always follows quench hardening of steel. Tempering consists of heating quenched, hardened steel in

Table 8–1. Stress relief treatments

Material	Treatment
Cast iron	900–1100°F, 0.5–5 hours
Carbon steel (cold-worked and heat-treated)	925–1200°F, 1 hour
Carbon steel weldment	1200°F, 1 hour

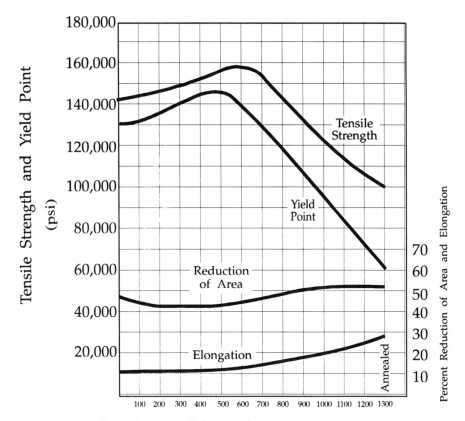

Figure 8–5. The effect of stress relief annealing treatments at various temperatures on cold-drawn steel bars.

the martensitic condition to some predetermined temperature between room temperature and the lower critical temperature, holding for an hour or so, and then cooling with air or oil. The rate of cooling from the tempering temperature usually has little effect.

Fully quenched hardened martensitic steel is generally very brittle, has high internal or residual stresses, and possesses low toughness and ductility. In this condition, steel has very few industrial applications. Tempering reduces brittleness of the hardened steel while still maintaining a high degree of hardness and strength (Figure 8–6). In general, tempering is used for the following reasons:

* Increase toughness
* Decrease hardness
* Relieve internal stresses
* Stabilize the structure
* Change the volume

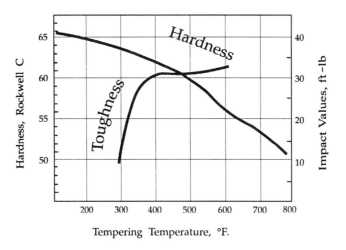

Figure 8–6. Effect of tempering temperature on the hardness of type 0–1 manganese nondeforming steel, oil quenches from 1475°F and tempered 1 hour.

Tempering operations are usually carried out at relatively low temperatures with few problems of surface scale formation. However, higher tempering temperatures may require atmospheric controlled furnaces or liquid baths to prevent surface scale formation. Air furnaces, controlled atmosphere furnaces, and liquid tempering baths are all popular methods for heating parts during tempering. Liquid baths containing oil, molten salts, or molten lead provide rapid heating, uniform temperature, and protection against oxidation. Oils may be used for heating temperatures in the 200–600°F range, salts for temperatures in the 350–2400°F range, and lead for temperatures in the 650–1600°F range. Although lead has a much higher heating rate than any of the salts, steel parts float atop molten lead and therefore require special fixtures to keep them submerged. Molten salts, on the other hand, tend to decompose and have various chemical effects on the metals, especially at high temperatures. Extreme caution must be observed to prevent water or wet parts from entering the tempering baths. Rapid steam formation can cause splattering or an explosion of the molten material.

It is common practice to divide tempering reactions into three stages:

Stage One During the first stage of tempering, martensite (formed in plain carbon steels) is heated to approximately 200°F. The martensitic structure is said to begin decomposing into low carbon martensite, and a transition carbide called the epsilon (ε) carbide, which has a hexagonal crystal structure, is formed. The composition of this phase is believed to lie close to $Fe_{2.4}C$. It contains approximately 8.4% carbon instead of the 6.7% carbon contained in cementite, Fe_3C. In this phase, the metal is also slightly harder than cementite. During this first stage of tempering, there is a slight reduction in the volume of the metal due to a shrinkage of no more than 0.0012 in./in.

Stage Two During the second stage of tempering, the temperature is increased to about 500° F. This softens the steel and changes it from austenite to bainite. The microstructure of the bainite consists of ferrite and the epsilon carbide. Bainite is somewhat similar to the decomposed martensite obtained in the first stage of tempering in that both contain epsilon carbide. The main difference, however, is in the shape of the matrix: In bainite, the matrix is cubic ferrite, whereas decomposed martensite has a matrix of low-carbon tetragonal martensite. Stage two of tempering is accompanied by an expansion of the workpiece, with dimensions growing as much as 0.010 in./in. due to the austenite–bainite transformation.

Stage Three During the third stage of tempering, which occurs rapidly between 500 and 675° F, the low-carbon martensite loses both its carbon and its shape; the carbon apparently diffuses to combine with epsilon carbide, forming very fine cementite particles. These carbide particles cannot be seen under the microscope, and the structure is still referred to as tempered martensite. It should be emphasized that tempered martensite does not take the form of plates characteristic of pearlite. Rather, a fine dispersion of somewhat spheroidal particles is formed. The particles' size depends on process time and temperature. There is a marked decrease in hardness during the third stage of tempering.

Summary Each stage of the tempering process affects the treated metal differently. Although tempering occurs in a series of distinct stages, the tempering curve shows a smooth decrease in hardness. Figure 8–7 plots the overall volume or length change from tempering for a particular oil hardening tool steel. In general, total growth during tempering (stages one, two, and three) is on the order of 0.1–0.3%. One application of this growth phenomenon would be to reharden worn or undersized reaming tools at a temperature that causes maximum growth, then regrind them to the proper diameter, thus extending their useful life.

Martempering

The martempering treatment (Figure 8–7) is carried out by heating the steel to the proper austenitizing temperature and then quenching it rapidly in a molten salt bath. The bath is held just above the M_s temperature (the temperature at which martensite begins to form). The steel is held long enough in the hot quenching bath to allow both surface and center to reach the same temperature. The

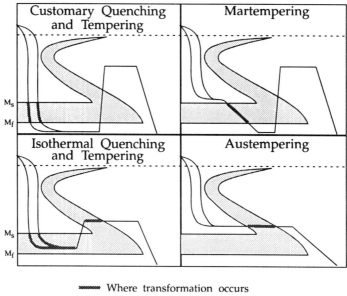

Where transformation occurs
M_S - Start of Martensitic Transformation
M_f - Finish of Martensitic Transformation

Figure 8–7. Schematic of the relationship of several tempering treatments to a typical IT diagram. Customary quenching and tempering, martempering, austempering and isothermal quenching.

steel is then removed from the quenching bath and cooled to room temperature. During the cooling from the M_s temperature to room temperature, the austenite changes to martensite. Since the temperature has been equalized throughout the steel, the transformation occurs uniformly, with minimal residual stresses and with greatly reduced danger of distortion and cracking. The heat treatment is completed by tempering the martensite to the hardness desired. The principal advantage of martempering is that fewer internal stresses result. This is because a lower temperature differential exists between the outer and inner portions of the steel workpiece during the transformation stage from austenite to martensite.

Austempering

The austempering cycle, shown schematically in Figure 8–7, is a hardening process based on transformation of austenite to bainite. From the isothermal transformation curve (Figure 8–8), it is apparent that during conventional quenching there is a strong likelihood of obtaining only two distinctive structures. The first structure, fine pearlite, occurs at the "nose" of the isothermal transformation diagram and has a Rockwell C hardness number between 40 and 42 (RC40–42).

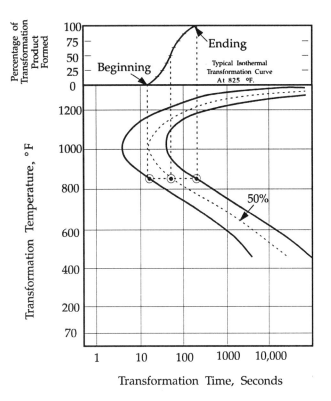

Figure 8–8. Isothermal cooling curve resulting from quenching a series of specimens into molten baths at various elevated temperatures.

The second structure, martensite, occurs at room temperature and has a Rockwell C hardness number between 60 and 65 (RC60–65). If one wishes to obtain a hardness of RC58, the customary heat treatment is to quench the steel to its maximum hardness (RC65) and then temper it to achieve RC58.

The austempering treatment can harden steel to RC58 or other hardnesses through isothermal transformation. The austempering cycle is similar to that of martempering, except that, with austempering, the steel is quenched to some selected temperature above the M_s and allowed to transform completely before cooling to room temperature. To obtain a hardness of RC58, austempering would be carried out at approximately 500°F for 1 hour. This would produce a structure of 100% bainite with uniform hardness.

Steel with a hardness of RC58 obtained by austempering is much tougher than a steel of the same composition and hardness obtained by other heat treatments. Figure 8–9 shows the relative properties of specimens hardened by conventional quenching and tempering and those hardened by austempering. The austempered specimens are much tougher and have much greater ductility than

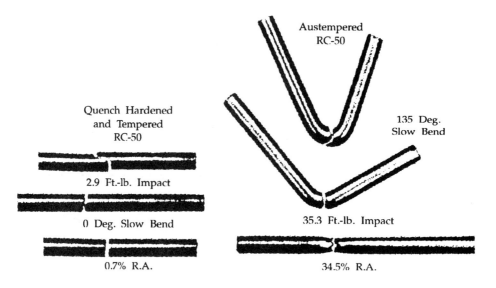

Figure 8–9. Improved toughness and ductility of austempered rods (*right*) compared to rods (*left*) hardened by conventional quench and tempering treatment (R.A. = reduction of cross-sectional area) (Edgar Bain Research Laboratory, U.S.X. Corp.).

quenched and tempered specimens. Since martensite is never produced during austempering, much of the danger of cracking is eliminated, and the amount of distortion or warping is greatly reduced.

Successful austempering depends on successful undercooling of austenite to the temperature of the hot quenching bath, without the formation of any softer forms of pearlite in the 900–1200°F range. This limits austempering to relatively thin specimens (generally less than 0.5 in. thickness) that can be rapidly cooled. Austempering has been successfully used on springs, lock-washers, screws, pins, needles, cultivator shovels, and similar parts.

Isothermal Quenching

The isothermal quenching cycle is shown schematically in Figure 8–7. Transformation is allowed to proceed to completion at a temperature only slightly above the M_s temperature. The constituents formed from this treatment are fairly hard. Increased toughness (and consequent reduction in hardness) can be achieved by raising the temperature; this also makes it unnecessary to quench the workpiece before tempering. Slow cooling to room temperature completes the cycle. The isothermal transformation curve for a particular steel provides a guide to the quench temperatures as well as indicating time requirements. Section size of the workpiece is very important. If the section is large, the cooling rate in the hot quenching bath will not be sufficient to cool the center of the part, and only partial hardening will occur.

8.4 Recrystallization Processes

In recrystallization processes, higher temperatures break down the crystal structure and allow for the formation of new grains. These processes, generally grouped together as annealing processes, are used to restore the soft condition of cold-worked metals. Annealing processes have three stages:

1. Heating to the proper annealing temperature
2. Holding or "soaking" at the annealing temperature
3. Controlled cooling from the annealing temperature

In annealing, slow, uniform heating of steel is desirable for two reasons. The first reason is to allow all the material in the workpiece to change phases at the same time so as to prevent unwanted stresses. When a workpiece is heated rapidly, its surface gets hotter more quickly than its center. This means that the surface reaches the critical temperature and changes into a different phase (austenite); the surface starts to shrink while the center is still expanding. This results in high tensile stresses at the surface and compressive stresses at the center of the workpiece. These internal stresses may be sufficient enough to produce warping or the initiation and propagation of fatigue cracks on the workpiece surface. Steels that are heated slowly have a more uniform temperature throughout. Thus, the surface and interior of the workpiece will both change to austenite at approximately the same time, resulting in minimum internal stress. The second reason for slow, uniform heating of a workpiece is that it allows for more uniform grain size. Rapid heating causes the edges and outer surface of a workpiece to have a rough structure while the grains in the center of the workpiece are still being refined. Rapid heating of large workpieces generally produces structures with nonuniform grain size. The proper temperature for annealing depends on the type of annealing process involved. The following recrystallization or annealing processes will be discussed: full annealing, process annealing, normalizing, and spheroidizing.

Full Annealing

Full annealing means heating hypoeutectoid steel (steel with a carbon content less than 0.85%) or eutectic steel (steel with a carbon content of 0.85%) to about

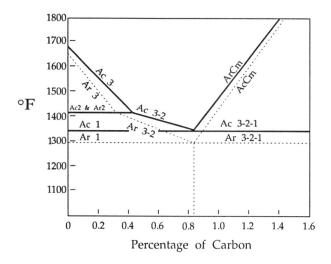

Figure 8–10. Critical temperature diagram showing normalizing, annealing spheroidizing, and hardening ranges for carbon steel.

100°F above the upper critical temperature, Ac_3 or Ac_{3-2}. Hypereutectoid steel (steel with a carbon content between 0.85% and 1.7%) needs to be heated 100°F above the Ac_{3-2-1} temperature for full annealing (Figure 8–10).

These temperatures produce maximum grain refinement with uniform, reoriented, and recrystallized, stress-free grains. Heating above the recommended temperature for full annealing results in large grains and causes hypereutectoid steels to have a coarse, brittle, cementite network surrounding the pearlite grains. The cementite network, or eggshell structure, in hypereutectoid steel is shown in Figure 8–11.

Process Annealing

Process annealing is used to restore ductility to cold-worked parts; it is often used in the sheet and wire industries to soften alloys for further cold forming and drawing. The process is normally carried out within the recrystallization temperature range (1000–1300°F). The intent of process annealing is to promote recrystallization, softening, and fine grains. For that reason, rapid heating and short holding time at temperature are used. Although using low temperatures reduces scaling, some surface oxide is formed.

Normalizing

Normalizing, another recrystallization heat treatment process, is similar to annealing in that the workpiece is heated to a temperature above the upper critical

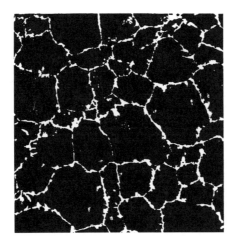

Figure 8–11. Cementite network surrounding pearlite grains in hypereutectoid steel, 100X.

range and held long enough to allow the structure to become completely austenitized. However, normalizing differs from annealing in that the rate of cooling is somewhat accelerated by allowing the steel to cool in air. The usual objectives of normalizing treatments are to secure a controlled, definite grain size and to produce a finer and stronger pearlite structure than is obtained by full annealing. Normalizing temperatures are approximately 100°F above the upper critical temperature. The undesirable structure resulting from hot forging is shown in Figure 8–12a; Figure 8–12b shows the normalized structure for the same steel.

Spheroidizing

Spheroidizing was designed to convert hard lamellar or network carbides of high-carbon steels into globular or spherical shapes in order to improve machinability or to obtain maximum workpiece ductility. The spheroidized type of structure (Figure 8–13) may be obtained by one of three methods:

1. Prolonged heating at a temperature just below the lower critical temperature, followed by relatively slow cooling

2. Alternately heating and cooling the workpiece to just above and just below the lower critical temperature

3. Short heating period at a high temperature (between 1380 and 1480°F), followed by relatively slow cooling

The first method, involving prolonged heating, is the slowest yet most frequently used spheroidizing process. It is popular because it does not require fur-

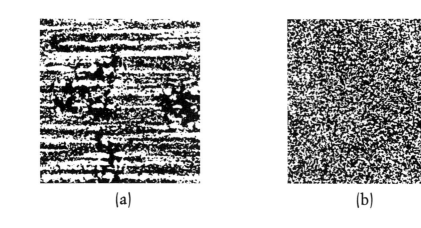

(a)   (b)

Figure 8–12. Grain structure before and after normalizing: (a) Undesirable microstructure of hot forged steel showing banded structure and mixed grain size, 75X. (b) Material after normalizing—undesirable structure has been corrected and grain size is uniform, 75X.

ther quenching or reheating operations. If the microstructure is initially coarse pearlite, the spheroidizing time may be 16 to 72 hours or longer at 1300°F. To facilitate spheroidizing, the initial structure should preferably be fine pearlite. Quenching the steel from the austenitic region often proves desirable prior to spheroidizing.

The second method, in which the temperature fluctuates above and below the critical temperature, can reduce spheroidizing time for small workpieces. The

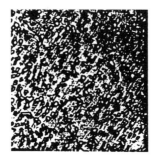

Figure 8–13. Spheroidized carbide microstructure of S.A.E. 3250 steel, suitable for automatic screw machine work, 750X, 3% Nital etch.

Table 8–2. Common annealing processes

Process	Heat treament	Resulting structure	Application
Full annealing (hypo and eutectic steels)	Heat to 50–100° F above A_{c3} or A_{c3-2}. Cool 50–100° F per hour.	Laminated pearlite	Improvement of low-carbon steels for deep drawing operations.
Process annealing	Heat to 1000–1300° F temperature range; follow with air cooling	Recrystallization of ferrite, distorted pearlite remains	Used as an intermediate anneal cold working of low-carbon steels including sheet metal and wire
Stress relieving	Heat uniformly to recommended temperature; follow with air cooling	No appreciable change from original structure	Used to relieve residual stress resulting from cold forming, casting, heavy maching, grinding, etc.
Normalizing carbon	Heat above A_{c3} or A_{cCm} then air cool	Fine pearlite	Used for plain carbon or low alloy steels. Used extensively for grain refinement of cast or welded steels
Spheroidizing (hypereutectic steels)	Heat to temperature between A_c and A_{cCm} followed by slow cooling through critical range. Hold just below critical temperature. Air cool	Spheroidized carbides in ferrite matrix	Improves machinability of high carbon steels, imparts maximum ductility and formability

length of time the steel is held at each temperature and the number of cycles through which it is heated and cooled depend on the carbon content and initial structure of the steel.

The third method, which involves higher temperatures, is used for tool steels and high alloy steels. Tool steels are generally spheroidized by heating to a temperature between 1380 and 1480°F for carbon steels. High alloy tool steels usually require higher temperatures. The steels are then held at temperature for 1 to 4 hours and then slowly cooled in a furnace. Spheroidized tool steels are much softer and more easily machined than annealed tools steel with coarse pearlitic structure.

8.5 Conclusion

Annealing processes are very useful in conditioning metals. They can be used to restore workpiece softness and ductility or to reduce the internal stresses of a workpiece without sacrificing significant material strength. Brittleness can also be reduced without sacrificing significant hardness. Softening processes are sometimes used to obtain a structure in the metal, such as spheroidite, so as to improve the work material's machinability.

Thermal Hardening Processes

9.1 Introduction

A metal workpiece may be hardened by a variety of methods and for a variety of reasons. Hardening processes usually modify the mechanical properties of a workpiece material. The changes in mechanical properties include increased wear resistance (due to increased hardness) and increased strength or resistance to deformation. Some physical properties, such as magnetic properties, also change as a result of being hardened.

Selecting a particular hardening process depends on the type of workpiece material and the degree of hardness desired. Some hardening processes are used to harden the surface only, and others are used to through harden the entire part (see Chapter 1 for a list of surface and through hardening processes). Surface coatings, mechanical work hardening, and thermal hardening processes are commonly used methods of hardening metals. Only thermal hardening processes will be discussed in this chapter.

Only certain metal alloys can be hardened by thermal hardening processes. In general, thermal hardening processes can be classified according to the following basic hardening mechanisms:

* Quench hardening of steel and cast iron
* Diffusion hardening of ferrous materials
* Age hardening of certain ferrous and nonferrous alloys

Quench hardening can be further divided into two subcategories, depending on the amount of carbon and other alloying elements present. In general, thermal hardening processes harden a workpiece material by the following steps:

1. The material is heated to a prescribed temperature
2. The material is held at that temperature for a prescribed time (referred to as *soaking time*)
3. Particles in the material diffuse at the surface and internally into a solid solution
4. The material is then quenched to prevent the diffused particles from escaping the lattice
5. The material is finally tempered or aged for a prescribed time to allow the particles to precipitate and yield the desired hardness and toughness

9.2 Basic Hardening Mechanism for Quench Hardening

Quench hardening processes are primarily used for through hardening. The basic mechanisms for thermal hardening of ferrous metals are as follows:

1. The ferrous metal is heated to 1330–1660°F, depending on the percentage of carbon in the iron or steel; at this point, a phase transformation occurs
2. Solid-state diffusion of alloying elements (carbon) occurs during the soaking time
3. A rapid quench causes a martensitic transformation of the iron or steel–carbon compound that distorts the lattice structure of the parent material, thus rendering it resistant to penetration, deformation, or abrasion
4. The quench is followed by a tempering process (at 400–800°F), during which time recovery of some diffused particles occurs, and the brittle martensitic structure becomes tougher while retaining strength and hardness.

These steps will now be discussed in more detail.

Phase Transformation

All metals are composed of basic building blocks called unit cells. A simple cell may be thought of as a small cube with atoms at each corner, as shown in Figure 9–1.

While there are many sizes and shapes of unit cells, two common types as-

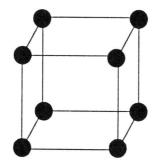

Figure 9–1. Simple cell.

sociated with iron and steel are the body-centered cubic (BCC) unit cell (Figure 9–2) and the face-centered cubic (FCC) unit cell (Figure 9–3).

At room temperature, iron exists as BCCs and has limited space for alloying elements between the interstices of the adjacent atoms. However, as iron is heated to 1660°F, it changes to a FCC. In this form, a small percentage of carbon atoms, which are much smaller than iron atoms, can fill the vacant spaces (Figure 9–4).

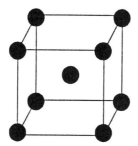

Figure 9–2. Body-centered cubic (BCC) unit cell.

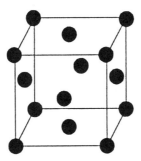

Figure 9–3. Face-centered cubic (FCC) unit cell.

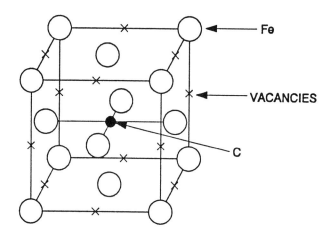

Figure 9–4. At 1660°F, iron cells change from BCC to FCC. A small percentage of carbon atoms fill the vacancies.

Solid-State Diffusion

Steel and other ferrous metals have a high concentration of alloying elements (mostly carbon) that exist microscopically in pockets. When heated, the concentrated alloying elements tend to diffuse uniformly throughout the workpiece material. This ability of alloying elements to migrate through the solid workpiece is called solid-state diffusion. Carbon atoms, for instance, can migrate from the workpiece surface toward the center at a rate of 0.020–0.040 in./hr. at temperatures ranging from 1550 to 1800°F.

In order for low-alloy steel to be hardened by heating and quenching, there must be between 0.45 and 0.95% carbon in the steel. If the steel is initially a low-carbon steel, the additional carbon may be added by heating the workpiece in a carbon-rich atmosphere (Figure 9–5).

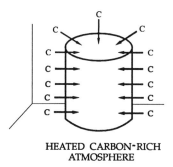

HEATED CARBON-RICH
ATMOSPHERE

Figure 9–5. Increasing the carbon content of the surface of a low-carbon steel through diffusion.

Such a carbon-rich atmosphere may be attained by the following:

1. *Pack carburizing*: packing the workpiece in a sealed container and heating it for several hours
2. *Gas carburizing*: heating the workpiece in a gaseous atmosphere composed of carbon monoxide
3. *Cyanide carburizing* or *cyaniding:* immersing the workpiece in a bath of molten cyanide salts.

These carburizing processes, however, are primarily used for surface hardening low-carbon steels, since it is difficult to diffuse carbon particles from the atmosphere to the center of the metal. Following the carbon enrichment process, the workpiece must be heated and quenched.

Quenching

As iron alloys that have been heated and soaked for a period of time are rapidly cooled (quenched) in water, brine, or oil, the FCC attempts to revert to its normal BCC. However, carbon atoms within the cell inhibit this transformation, and a distorted cell results, with one axis longer than the other two. This distorted cell is called a tetragonal unit cell. Excess carbon atoms also combine with iron to form a hard iron carbide, Fe_3C, which appears in the microscope as a pole of strain consisting of sharp crystals known as martensite. The quenching of heated iron in which carbon has diffused results in the creation of a hardened steel workpiece.

Tempering

A fully quench-hardened steel workpiece is quite brittle. However, if hardened steel is gradually heated to about 1000°F, the sharp martensitic structure is replaced with a fine structure composed of alternate platelets of iron and iron carbide known as pearlite. Usually, hardened steels are tempered at 400–800°F. This results in an intermediate structure that is still hard, but not brittle.

9.3 Basic Hardening Mechanism for Diffusion Hardening

Diffusion hardening processes are used for surface hardening. One mechanism for producing thin, hard cases on steels, *nitriding,* involves solid-state diffusion of nitrogen instead of carbon. In this process, a workpiece is held at around 1000°F for several hours; during this time, very hard iron nitrides form in a shallow case on the surface of the workpiece. This thin case resists softening at elevated temperatures and is used when very hard, wear-resistant surfaces are desired. In *carbonitriding*, nitrogen gas is bubbled into a molten cyanide salt bath, and the resultant workpiece surface is composed of both carbides and nitrides.

9.4 Basic Hardening Mechanism for Age Hardening

Age hardening is primarily a through hardening process. Certain nonferrous alloys of aluminum, copper, and nickel, as well as some ferrous alloys, may be hardened by the mechanism known as *precipitation hardening* or *age hardening*. In this process, the workpiece is heated and allowed to soak for an appropriate period of time. The workpiece is then quenched to near room temperature. Upon quenching, the workpiece is relatively soft. After aging, precipitates begin to form, causing a distortion of the inner structure of the workpiece; the workpiece becomes hardened. A special copper–beryllium–nickel alloy containing 2% beryllium can be hardened to as high as RC54 by this process. Nonsparking tools, such as wrenches, chisels, and hammers, may be hardened by this process, as well as high strength nickel–aluminum alloys for highly stressed parts that must operate in corrosive environments.

Workpiece Materials/Process Conditions

The workpiece materials in Table 9–1 can be subjected to age hardening treatments. Process conditions of time and temperature for solution treatment and for age hardening are given. The alloy is heated to the temperature shown, after which it is quenched in water and then aged at the specified temperature for the specified length of time.

9.5 Basic Mechanism for Induction Hardening

Induction hardening is primarily a surface hardening process, but through hardening is possible, depending on workpiece thickness and equipment capabilities. A ferrous metal workpiece with medium- to high-carbon content is heated by a high frequency electromagnetic field and rapidly quenched to produce the har-

Table 9-1. Process conditions for age hardening treatments

Material	Designation	Solution treatment		Precipitation treatment	
		Temp. (℉)	Time (Hrs)	Temp. (℉)	Time (Hr)
Aluminum	2024	920	0.25–4	375	7–9
Aluminum	6061	920	0.25–4	350	6–10
Aluminum	7075	870	0.25–4	250	24–28
Aluminum–Bronze	Cu 90–Al 10	1650	1	700–1100	24
Beryllium–Bronze	Cu 97, Be 2.5, Co 0.5	1975	1	475–575	24
Special	Cu 56–52 Mn 22–24 Ni 22–24	1200	1	660–840	24
K-Monel	Ni 65, Cu 30 Al 2.8, Ti 0.5	1600	0.5	1100	8–16

dened structure. The workpiece is placed inside a coil and subjected to a high frequency alternating current. A rapidly changing magnetic field induces eddy currents in the workpiece surface, which is heated by electrical resistance (Figure 9–6).

With this process, selected workpiece surfaces, such as the teeth on a gear, may be hardened for wear resistance, while other surfaces can remain soft for subsequent machining or where toughness properties of the workpiece material are to be maintained. The localized heating can also minimize distortion and maintain strict tolerances. The process can be customized for special shapes and is fairly versatile. It offers great potential for maintaining production at reasonable costs, and it can be automated. When a part is induction hardened, the surface stresses of the part are left in compression; this can increase the fatigue strength of the part. As mentioned, this process is limited to ferrous metal parts.

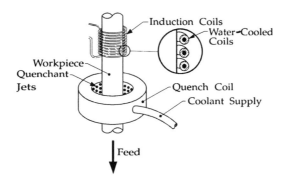

Figure 9–6. Induction hardening process.

9.6 Common Heat Treatment Problems

Table 9–2 shows common problems encountered during the heat treatment of steels, along with possible causes and remedies.

Table 9–2. Heat treatment of steels

Problems	Possible causes	Remedy
Warping	Nonuniform quenching practice	Employ spray or agitated quench
	Improper support during heating	Support with brick, cast iron chips, or spent coke
	Release of machining stresses	Machine equal amounts from surface of part or anneal prior to heat treatment
	Unbalanced design	Clamp in fixture designated to balance mass
	Failure to strain relieve prior to treatment	Strain relieve
Dimensional changes	Release of stresses from previous cold working	Strain relieve prior to hardening
	Unpredicted thermal stresses	Balance mass with quench fixture
	Severe quenching practice	Change to less severe quenching media or warm quench bath
	Failure to temper or stabilize properly	Employ stabilizing or subzero treatment
	Dimensional changes for some are predictable and normal	Use table listings to predict size change
	Transformation of retained austentite	Employ multiple tempers or subzero treatment
	Overheating or underheating	Check furnace control and recommended temperatures
Cracking	Failure to temper immediately after quenching	Temper before it reaches room temperature (approximately 150°F)
	Improper quenching medium	Use less severe quench
	Poor design; e.g., sharp corners, unbalanced mass	Discuss with designer
	Failure to preheat properly	Preheat as recommended
Failure to harden	Quench not drastic enough	Employ more drastic quench
	Hardening temperature too low or nonuniform heating	Check recommended temperature
	Mislabeled steel	Make test run on sample, or get it analyzed
	Severe decarburization	Use controlled atmosphere or liquid heating bath
	Tempering temperature too high	Use recommended temperature

Table 9–2. Heat treatment of steels *(continued)*

Problems	Possible causes	Remedy
Soft spots	Decarburized case	Use controlled atmosphere or liquid heating bath
	Excessive heat treat scale	Use controlled atmosphere or liquid heating bath
	Quench bath too hot	Check temperature
	Improper agitation	Review recommended procedure
	Contaminated quenching bath	Clean, filter, or change
Excessive brittleness	Improper quenching medium	Use recommended quench
	Failure to temper	Temper immediately after hardening
	Excessive hardening temperature	Use recommended temperature
	Coarse grain size	Use recommended temperature
	Mechanical stress raisers; e.g., sharp corners	Discuss with designer or use air hardening steel

Surface Preparation Processes

10.1 Introduction

Surface preparation is a subfamily of surface finishing on the taxonomy chart (Chapter 1). The purpose of surface preparation processes is to remove contaminants and other undesirable elements from the workpiece surface so as to prepare it for further processing. Surface preparation processes are classified into three categories: descaling, deburring, and degreasing. Descaling processes remove dirt and scale (a flaky oxide layer on metals). Deburring processes remove burrs from the workpiece and polish its surface. Degreasing removes oil and grease. Within each of these categories there are mechanical, thermal, and chemical processes.

Surface preparation processes can be used on a wide variety of materials. Most metals, as well as glass, plastics, ceramics, and other materials can be processed by surface preparation operations.

10.2 Descaling

Process Mechanics

Descaling processes remove dirt and scale mechanically, thermally, or chemically. The process mechanics are different in each case. Examples of descaling processes are wire brushing and pickling (Figure 10–1).

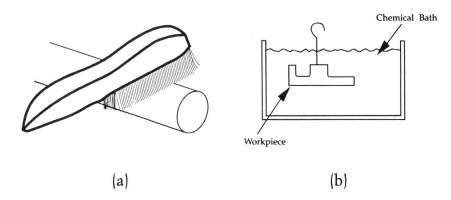

(a)   (b)

Figure 10–1. Descaling processes: (a) Wire brushing (mechanical descaling) and (b) pickling (chemical descaling).

In *mechanical descaling,* the scale is removed by abrasive particles, elastic tooling, or rigid tooling. Abrasive particles can be applied through a jet stream of air or liquid. Rapidly moving particles strike the surface and chip away the layer of scale. This is the same principle that is used in abrasion processes in the mechanical mass-removal family. In elastic tooling, wire, fiber, or other brushes are used to flake away the contaminants. In rigid tooling, belts and wheels that contain abrasive particles erode away the scale. Sometimes conventional mechanical mass-reducing processes such as grinding or belt sanding are used to remove scale. In *thermal descaling,* better known as flame cleaning, a high velocity oxyacetylene flame cleans the surface of steel and iron. The highly oxidizing flame burns off the scale. This process can also be used to remove paint. In *chemical descaling,* better known as pickling, a workpiece is dipped into an acid solution and then washed. The solution chemically loosens and breaks down the scale. Although this method effectively removes scale, it does not remove dirt well, so the workpiece should be cleaned before pickling.

Process Conditions

Each process has specific parameters that determine the outcome of the operation. The following general parameters need to be considered:

Mechanical Descaling
   * Type of tooling
   * Abrasive type and size
   * Type of binder (in elastic and rigid tools)

* Speed and feed
* Relative motions between tool and workpiece

Thermal Descaling
* Flame temperature
* Type of workpiece material

Chemical Descaling
* Bath temperature
* Concentration of alkaline solution
* pH of solution
* Purity of solution
* Workpiece material (chemical resistance)
* Immersion time
* Thickness of layer of scale

Characteristics of Descaling Processes

Properly descaled workpieces have oxide-free surfaces. Descaling processes should be used when the oxides need to be removed for another process. Often, descaling operations must be followed immediately by another process to protect the workpiece surface; otherwise, oxidation will occur. Care should be taken with all chemical processes to prevent injury to operators and harm to the environment.

10.3 Deburring

Process Mechanics

Process mechanics for deburring processes also vary greatly depending on the specific process used. Deburring processes include barrel tumbling and electrochemical deburring (Figure 10–2).

In *mechanical deburring*, burrs are removed when abrasive chips suspended in a slurry are applied to the workpiece by tumbling, vibration, or some other method. *Thermal deburring* is achieved by a chemical reaction that is facilitated by the application of heat (thermochemical deburring). In *chemical deburring*, the workpiece is dipped into a solvent bath, and a chemical reaction that is driven by electric current dissolves away the burrs. These thermal and chemical deburring processes are similar to etching or electrochemical milling processes.

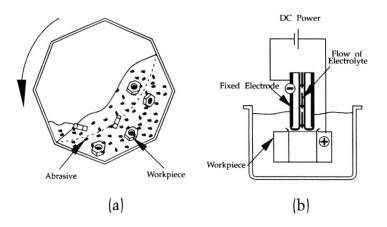

Figure 10–2. Deburring processes: (a) Barrel tumbling (mechanical deburring) and (b) electrochemical deburring.

Process Conditions

Each process has specific parameters. The following general parameters should be regulated to ensure good results:

Mechanical Processes
   * Type of tooling
   * Type and size of abrasive particles
   * Type of binder (holding the particles)
   * Speed and feed rates
   * Pressure at which tooling is applied
   * Relative motions between tooling and workpiece

Thermal and Chemical Processes
   * Degree of agitation
   * Bath temperature
   * Concentration and type of solvent or electrolyte
   * pH value
   * Purity of solvent
   * Workpiece material (chemical resistance)
   * Immersion time

Characteristics of Deburring Processes

Deburring processes leave the workpiece smooth and free from burrs. Usually, only a small amount of the workpiece is removed, so tight tolerances can be maintained. Due to the nature of these processes, sharp corners and other intricate geometries may be lost.

10.4 Degreasing

Process Mechanics

Grease, oil, and other contaminants can be removed either mechanically or chemically. Ultrasonic degreasing is an example of a mechanical degreasing process, and vapor degreasing is an example of a chemical degreasing process (Figure 10–3). In *mechanical degreasing* (or ultrasonic degreasing), cavitation in a liquid bath is created by ultrasonic vibration. Cavitation, the creation of small vacuum areas as a liquid is pulled apart, helps break up and remove contaminants from a workpiece surface. *Chemical degreasing* involves dipping the workpiece in a chemical bath to remove surface oil and grease. Two basic chemical processes can occur—emulsification and dissolution. In emulsification, the contaminant is loosened, lifted, and clustered into tiny spheres but is not necessarily removed from the surface. An additional washing process is necessary to remove the contaminant after emulsification. Typical emulsions are alkaline solvents (such as caustic soda), chlorinated hydrocarbons (such as trichlorethylene), soaps, and alcohols. In dissolution processes, the contaminant is chemically broken down (dissolved) and carried away in solution. Dissolution solvents are usually acids. Sometimes

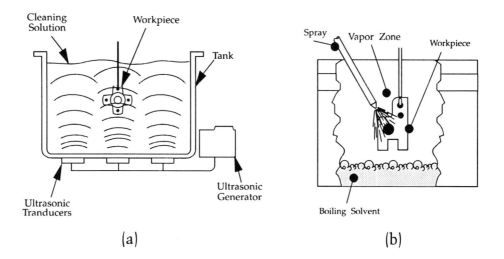

Figure 10–3. Degreasing processes: (a) Ultrasonic degreasing (mechanical degreasing) and (b) vapor degreasing (chemical degreasing).

dissolution processes require an additional operation to remove the solvent from the workpiece.

Process Conditions

General parameters for mechanical and chemical degreasing processes include the following:

Mechanical Processes
* Frequency of vibration
* Immersion time

Chemical Processes
* Bath temperature
* Concentration and type of solvent
* pH value
* Purity of solvent
* Workpiece material (chemical resistance)
* Immersion time
* Type of contaminant

In chemical processes, it is important to select a suitable solvent for the type of workpiece material being processed and the type of grease or oil being removed. Some solvents will not dissolve grease, whereas other solvents are strong enough to dissolve many workpiece materials.

Characteristics of Degreasing Processes

Degreasing operations effectively remove grease and oil from a workpiece surface. These processes are often used prior to other surface finishing operations. Degreasing is sometimes even used before some of the other surface preparation processes mentioned in this chapter. Workpieces are left free of grease and oil without any change to shape or dimension.

Surface Coating Processes

11.1 Introduction

Surface coating is in the surface finishing family along with surface preparation and surface modification. The purposes of surface coating are as follows:

* To supply the workpiece with certain functional properties such as hardness, wear resistance, and corrosion resistance
* To give the workpiece an attractive appearance

The capabilities of coating processes have made them increasingly important in industry. Surface coating technology allows a manufacturer to select a lower grade material and then give it certain desirable properties. Surface technology can create properties that could not be created otherwise.

11.2 Basic Processes

The surface coating family is divided into three groups: mechanical, thermal, and chemical coatings. The coating processes are classified into these three groups based on how the coating is prepared and applied.

11.3 Mechanical Coating Processes

Process Mechanics

Mechanical coating processes are characterized by the following:

* Bonding forces between the coating and the workpiece are primarily adhesive
* The coating material is prepared mechanically (e.g., stirring or mixing) or thermally (e.g., melting or evaporation)
* The coating is applied by dipping, spraying, electrostatic forces, or other mechanical means
* Workpiece materials may be metallic or nonmetallic
* Coating materials may be metallic or nonmetallic
* A workpiece surface is covered with coating material by a suitable motion between the workpiece and coating applicator tool

Figure 11–1 shows two examples of coating processes. In Figure 11–1a, a dip

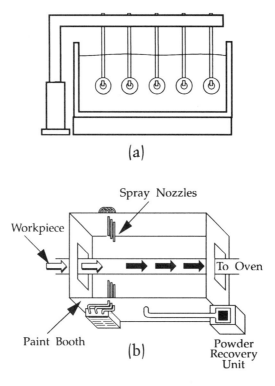

Figure 11–1. Mechanical coating processes: (a) Dip coating and (b) electrostatic painting.

coating process, the workpiece to be coated is dipped in the liquid coating material. Figure 11–1b shows an electrostatic painting system in which charged powder particles are guided into an electrostatic field that is formed between the workpiece and an electrode. After application, thermal treatment and curing are necessary to finalize the process. Another coating process that is not shown here consists of tumbling the workpiece together with the coating material in powdered form. This process is called *mechanical plating*.

Process Conditions

General parameters that influence mechanical coating processes include the following:

* Cleanliness of the workpiece surface
* Shape and size of the workpiece
* Type of workpiece material

Besides the basic considerations given above, each process has specific parameters that should be followed for optimal results. Some of the specific parameters of mechanical coating processes are given below.

Dipping Processes Type of coating, coating viscosity, temperature of dipping bath, amount of contamination in dipping bath, workpiece geometry, and type of workpiece material

Spraying Type of coating, coating melting point, size of coating droplets, velocity of droplets, coating temperature, temperature of workpiece, temperature of atmosphere, relative motion between coating applicator and workpiece, workpiece shape, and type of workpiece material

Electrostatic Painting Type of coating, electrostatic field, workpiece shape, and relative motions between applicator tool and workpiece; the workpiece must also be electrically conductive

Mechanical Plating Type of coating, size of coating particles, relative motions between coating particles and workpiece (including "hammering" action that occurs in some processes such as tumbling)

This list is not sufficiently detailed for direct use. Its purpose is to indicate that several parameters must be controlled to achieve good results.

Coating and Workpiece Materials

In general, coating materials should meet the following criteria:

* Be able to bond adhesively to the workpiece
* Meet surface property requirements of hardness, corrosion resistance, wear resistance, and appearance
* Be easy to apply

For mechanically applied coatings, the most common coating materials are paint, lacquer, enamel, plastics, and metals (including aluminum, zinc, tin, and tin–lead alloys). Other mechanical coating processes use coating materials made from high temperature resistant metals such as tungsten, cobalt, chromium, and steel. Refractory ceramics such as aluminum oxide and zirconium oxide are also useful coating materials. Many types of workpiece materials can be coated with mechanical coating processes. The following can be said about workpieces to be coated:

* They must be dry and clean to ensure good adhesive bonding
* For some processes, the workpiece must be electrically conductive
* Workpieces made of any metal and most nonmetals may be coated

Many materials can be coated and used as coatings in mechanical coating processes.

11.4 Thermal Coating Processes

Process Mechanics

In thermal coating processes, heat is used to melt or vaporize the coating material, which is then deposited on the workpiece. Thermal coating processes are divided into several subcategories; we will discuss flame spraying processes and vaporized metal coating processes. The process mechanics are slightly different for these subcategories. In flame spraying processes, metals or ceramics are melted, and the molten droplets are atomized and applied with a jet of compressed air (Figure 11–2a). The coating materials can be melted by gas flame, electric arc, or plasma arc. The bond between the coating and the workpiece is based on adhesion. In vaporized metal coating processes, coating materials are vaporized by electric current or applied heat and then deposited on the workpiece (Figure 11–2b). There is usually a chemical bond between the workpiece surface and the coating material.

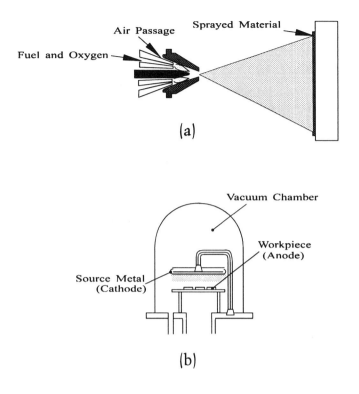

Figure 11–2. Thermal coating processes: (a) Flame spraying process (plasma spraying) and (b) vaporized metal coating process (sputtering).

Process Conditions

The main parameters for flame spraying processes and vapor metal coating processes are listed below:

Flame Spraying Processes
  * Workpiece surface cleanliness
  * Workpiece surface roughness

Vapor Metal Coating Processes
  * Type of coating material
  * Original form of the coating metal
  * Evaporation temperature
  * Exposure time

In flame spraying processes, it is important to keep the workpiece free from moisture and oil. Surface roughness is also important because the coating material is better able to adhere to a rough surface. In vaporized metal coating processes, the type of coating material, the evaporation temperature, and the exposure time are important issues. Several different metals are used in vapor coating processes, and each coating metal requires a different evaporation temperature to coat the workpiece successfully. Coating metals come in the form of wire staples, powder, wire, or pellets; the form of the metal used usually depends on the type of metal used.

Coating and Workpiece Materials

A wide variety of materials can be coated and act as coating materials. In flame spraying processes, metals or ceramics are melted to be used as coating materials. Typical coating metals include zinc, aluminum, and tungsten. The metals that can be coated with this process include steels, aluminum alloys, and nickel. In vapor metal coating processes, many metals, metal alloys, and metal compounds can be used as coatings. Typical workpiece materials are metals, but occasionally plastic, glass, and paper workpieces are coated.

11.5 Chemical Coating Processes

Process Mechanics

In chemical coating processes, the coating is applied chemically. Chemical coating processes have the following characteristics:

* Bonding is based on adhesive forces
* Application is a chemical process
* Coating is applied by chemical deposition, oxidation, or some other means
* Sometimes the chemical deposition of the coating is driven electrolytically
* Application of the coating is usually followed by rinsing, drying, curing, or a combination of these
* Coating materials are metallic; they can be pure metals, alloys, or metals with dispersed particles of other substances (such as oxides or teflon)

Chemical coating processes include the coating process itself as well as any postoperations required such as rinsing, drying, and curing.

Chemical coating processes are sometimes divided into two subgroups: electroplating processes and chemical conversion processes. These processes have unique characteristics. In electroplating processes, deposition of the coating is controlled by an electric field, which prohibits deposition of the coating in holes and creates an uneven distribution of the coating material. In chemical deposition

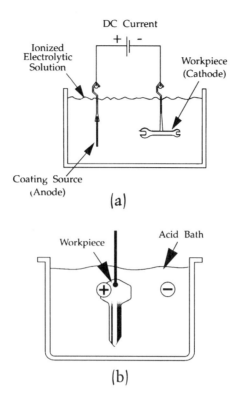

Figure 11–3. Chemical coating processes: (a) Electroplating and (b) anodizing.

processes, the coating material is uniformly distributed and can be used to coat the inside of holes. Figure 11–3 shows the processes of electroplating (or electrolytic deposition) and chemical conversion (or more specifically, anodizing).

Anodizing was developed especially for aluminum. Anodizing provides a conversion coating based on the oxidation of the workpiece surface by use of chromic acid, sulfuric acid, or another powerful oxidizing agent. In anodizing, organic coloring may be added to the workpiece, allowing many colors to be produced. For materials besides aluminum, other chemical conversion processes have been developed (such as chromate conversion and phosphate conversion of steels).

Process Conditions

The concerns in coating processes are different, depending on the specific process. The main parameters that influence the electroplating processes include the following:

* Surface cleanliness
* Type and concentration of electrolyte

* Current density
* Bath temperature
* Workpiece shape
* Arrangement of workpiece in the bath (the setup)

For chemical conversion processes, the following parameters influence the results:

* Surface cleanliness
* Type of bath (constituents and concentrations)
* Bath temperature
* Arrangement of workpiece in the bath (the setup)

In chemical coating processes, workpiece surfaces must be absolutely clean. Sometimes etching is used to prepare the workpiece to be coated. With electroplating processes, sometimes the workpiece must be made electrically conductive before the plating can take place.

Coating and Workpiece Materials

In general, coating materials should have the following characteristics:

* Soluble in a suitable solvent
* Electrically conductive (for electroplating processes)
* Reducible from the solution (in chemical plating processes)
* Able to create the desired bonding
* Able to create the desired properties

Examples of coating materials include tin, zinc, lead, cadmium (prohibited by law in some countries), chromium, copper, nickel, aluminum, steel, gold, platinum, silver, and various metal alloys. Typical materials to be coated include copper, brass, nickel, nickel-based alloys, aluminum, steel, nonmetals, and certain plastics.

11.6 Characteristics of Coated Components

Coated components have properties such as wear resistance, corrosion resistance, hardness, high temperature resistance, weldability, and attractive appearance. Typical coating thicknesses range from 0.0005 to 0.025 in. Coating processes allow for the use of lower grade materials, since the desired properties can be obtained by applying the appropriate coating.

To test the coatings, many standardized methods such as the salt-spray test and the wear test have been developed. Coating is a relatively new and promising field.

Index